GEORGE PLACZEK

A NUCLEAR PHYSICIST'S ODYSSEY

GEORGE PLACZEK

A NUCLEAR PHYSICIST'S ODYSSEY

ALEŠ GOTTVALD
Academy of Sciences, Czech Republic

MIKHAIL SHIFMAN
University of Minnesota, USA

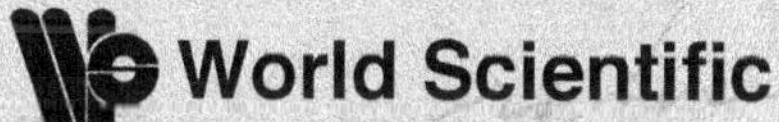

World Scientific

NEW JERSEY · LONDON · SINGAPORE · BEIJING · SHANGHAI · HONG KONG · TAIPEI · CHENNAI · TOKYO

Published by

World Scientific Publishing Co. Pte. Ltd.

5 Toh Tuck Link, Singapore 596224

USA office: 27 Warren Street, Suite 401-402, Hackensack, NJ 07601

UK office: 57 Shelton Street, Covent Garden, London WC2H 9HE

Library of Congress Cataloging-in-Publication Data
Names: Shifman, Misha, author. | Gottvald, Aleš, 1954–   author.
Title: George Placzek : a nuclear physicist's odyssey / Misha Shifman (University of Minnesota, USA),
     Aleš Gottvald (Academy of Sciences, Czech Republic).
Description: Singapore ; Hackensack, NJ : World Scientific Publishing Co. Pte. Ltd., [2018] |
     Includes bibliographical references and index.
Identifiers: LCCN 2018018791| ISBN 9789813236912 (hardcover ; alk. paper) |
     ISBN 9813236914 (hardcover ; alk. paper)
Subjects: LCSH: Placzek, G. (George), 1905-1955. | Nuclear physicists--Czech Republic--Biography.
Classification: LCC QC774.P53 S55 2018 | DDC 530.092 [B]--dc23
LC record available at https://lccn.loc.gov/2018018791

British Library Cataloguing-in-Publication Data
A catalogue record for this book is available from the British Library.

Cover and graphic design by Polina Tylevich

First published 2018 (Hardcover)
Reprinted 2020 (in paperback edition)
ISBN 978-981-122-134-7 (pbk)

For any available supplementary material, please visit
https://www.worldscientific.com/worldscibooks/10.1142/10900#t=suppl

Does human nature undergo a true change in the cauldron of totalitarian violence? Does man lose his innate yearning for freedom? The fate of both man and the totalitarian State depends on the answer to this question. If human nature does change, then the eternal and worldwide triumph of the dictatorial State is assured; if his yearning for freedom remains constant, then the totalitarian State is doomed.

Man does not renounce freedom voluntarily. This conclusion holds out hope for our time, hope for the future.

Vasily Grossman

Life and Fate

Late photograph of Georg Placzek. Courtesy of AIP Emilio Segrè Visual Archives.

Foreword

Mikhail Shifman

...Whenever one met [George Placzek] his deep insight in scientific and human problems was a source of inspiration and encouragement.

Niels Bohr
December 31, 1955[1]

Placzek's name doesn't come-up all that often now, but he was highly thought of then. He briefly succeeded Hans Bethe as leader of the Theoretical Division. He left shortly after his promotion due to health issues...

Alan Carr,
the LANL Historian and Archivist
September 2016

Unlike other trailblazers of nuclear physics, such as Bethe, Frisch, Peierls, Weisskopf, ..., George Placzek[2] did not leave his recollections or life story notes; moreover, his epistolary legacy as we know it now is quite scarce. His life and fate is known to us mostly from memoirs of his friends and colleagues and a few documents which survived in various archives.

Hans Bethe, his close friend, once characterized[3] Placzek as follows:

> Placzek had the highest standards of anybody I have ever known, both in ethics and in science, and his high standards in science were very detrimental to his productivity. He never was satisfied with anything he himself did.

---

[1] See page 221.

[2] His name at birth was Georg Placzek (the first language in their family was German). In France Georg Placzek spelled his first name as Georges, and in England, Canada and the United States as George. This latter version is widely accepted in the historic literature.

[3] This citation is taken from Silvan S. Schweber, *Nuclear Forces: The Making of the Physicist Hans Bethe*, (Harvard University Press, 2012.)

I was writing my one and two electron problems, and he was writing an article for another encyclopedia, Marx's *Handbuch der Radiologie*, in which he finally produced an article which has been used quite a lot. But it was certainly a very difficult birth. He would have a piece of paper in front of him, and then in the middle of the piece of paper somewhere, he would write something in very large letters, and that was all. Then he would disappear into the library and investigate the history of the scattering of light, going back to the early 1800s, and the library in Rome was especially good at that. I'm surprised it didn't contain any Roman books. Placzek would delve for the whole day then into some paper of 1830 or so, in which somebody had done something about the scattering of light, all as part of his article about scattering of light: Rayleigh scattering and the Raman effect. And he wanted everything proved very precisely in his own work. After having filled a few sheets of paper, he would then talk to me at dinner, and say, "I have again discovered a scientific paradox," and he would explain that over dinner. We would then try to resolve it, sometimes successfully, and then he would continue the next day. And so, his article progressed very slowly, and he constantly said, "Well, someday Mr. Marx will come here" – Marx was the editor of that Encyclopedia – "and somehow tell me to deliver the manuscript immediately." I think the manuscript was overdue already by a year or two, and in fact, at the end when Placzek had gone to Copenhagen, after Rome, Mr. Marx, or his emissary, did come, and took whatever Placzek had written, which was very incomplete, and said "well, never mind, that it is incomplete, I'll take what you have," which made Placzek very unhappy. Then somehow he managed, in proof, to add the second half. This is very characteristic of Placzek.

Needless to say, I never personally met Placzek for the simple reason that I was born approximately at the time when Placzek died. But I know quite well other physicists with a similar mentality. For instance, in 2002, I wrote[4] about Arkady Vainshtein, my co-author of thirty

---

[4]See https://www.academia.edu/17865243/

years:

> Arkady is a deep thinker. He is the deepest thinker of all people I am closely acquainted with. When he gets seriously interested in a certain physics problem – let us call it "problem A" – his mind sends a powerful urge to start digging. The outside world ceases to exist, the work continues almost on the 24/7 basis. A sophisticated phantasmagoric construction gradually emerges in Arkady's mind. Being left to himself, he would never return back. The problem A would lead to a set of subproblems $a1$, $a2$, and so on, which, in turn, would continuously evolve into a set of sub-subproblems α_{11}, α_{12}, ..., α_{21}, α_{22}, etc. Let alone related problems B, C, D,... The fractal nature of such an approach requires from Arkady a non-commensurate amount of time and effort. A little baroque exercise at level α whose impact on the general picture is minute, is as important to him as everything else. It may take weeks or months. Nevermind. Being left to himself, Arkady would never say: "This is the answer, I pause here to let other people know of what I have achieved." For him, the pleasure of finding out how things work is sufficient by itself. You may call him superperfectionist. Yes, that's the right word, extreme perfectionist. Only strong external impulses can extract him from the deepening fractal structure of his making. The onset of the vacation season may serve as such an impulse. Another option is to distract him by suggesting a new and more challenging problem. In this latter case the attraction of the new problem must be overwhelming, to overcome the inertia of the original motion. Upon forced return from the n-th intellectual journey, nothing can be taken for granted with Arkady. Even a solid baggage of results and insights acquired *en route* is no guarantee that the corresponding paper will ever see the light of the day. To make a decision to start writing a paper is a torture for Arkady. Even more so the process of writing. Every research project, its merits notwithstanding, has loose ends and dark corners. At the discussion

---

Arkady_Vainshtein_40_Year_Journey_in_Theoretical_Physics.

> stage everything is volatile, up in the air. What was a
> loose end today might find a perfect match tomorrow.
> But when you put this on paper, this is it. Every string
> of Arkady's superperfectionist ego protests. The neces-
> sity to document things before they are fully complete
> (and they never are) burns Arkady out.

I know for a fact that such people tend to generously share their ideas with friends and colleagues, rather than to publish them immediately, and therefore their impact on the community is much stronger than one could judge from their publications. Indeed, as we will see later, there is ample evidence that this was exactly the case with George Placzek. Many of his colleagues recall that his sharp remarks and comments were instrumental in their studies or the very inceptions of research topics.

In the nascent nuclear physics community Placzek was very early recognized as one of the major players. Thus, on February 9, 1932, in a private letter to his wife, Rudolf Peierls wrote from Leipzig:

> The conference here is incredibly interesting because al-
> most all good physicists came: Kramers, Ehrenfest, de
> Haas, Kronig, Kapitza, Becker, Gerlach, Placzek, Teller,
> Møller, Bethe, and so on.[5]

For reasons I do not quite understand, historians of science by and large kept silence about Placzek, despite his important role in the development of nuclear physics in the 1930s and in the Manhattan Project in the 1940s. By the way, while his name was almost forgotten for half a century in the West, it was always alive in the USSR. Since I came to ITEP[6] in 1970 I heard various stories and jokes about Placzek more than once. Apparently, their origin can be traced back to Lev Landau and, in part, to Alexander Akhiezer, who worked with Placzek in Kharkov in the 1930s. They were passed to us through their students. *"Placzek, Placzek, dai Kalatchik..."*[7]

The absence of historical memory with regards to Placzek was only recently remedied, to an extent, by Aleš Gottvald and Jan Fischer. Jan Fischer became interested in Placzek's legacy, closed some gaps in his biography and, after half-century of oblivion, published in the *CERN Courier* issue of August 23, 2005, an article entitled "Georg Placzek

---

[5] Translation from Russian is mine. -MS

[6] Institute for Theoretical and Experimental Physics in Moscow.

[7] In Russian *dai* means give. It is pronounced like "shy" with the replacement of sh by d. See page 289.

(1905-1955)." This article was dedicated to the centennial anniversary of Placzek's birth and attracted certain attention in the physics community. Among other sources, Jan Fischer quoted Aleš Gottvald's publication (in Czech) in the journal *Československý časopis pro fyziku*.[8] This was the first rather detailed biography of Georg Placzek. Meticulous research carried out by Aleš Gottvald before 2005 allowed him to establish many new facts about Placzek's life, his family roots, his role in the Manhattan project, and his tragic death. An expanded and revised English version of this article is presented in this Volume on pages 5-60.[9]

On September 21-24, 2005, an International Symposium to commemorate Placzek's 100th anniversary was held in Brno, Czech Republic.[10] World experts in the areas of Placzek's research and some historians of science descended on Brno to discuss Placzek's life and work. This event attracted attention outside scientific community: it was covered by Czech mass media.

I think it would be in order to quote here an excerpt from the Symposium announcement.

> Two life anniversaries of George Placzek will be commemorated in 2005: that of his birth on September 26, 1905, and that of his passing away on October 9, 1955. Placzek was an outstanding scientist who made substantial contributions to the fields of molecular physics, scattering of light from liquids and gases, the theory of atomic nucleus and the interaction of neutrons with condensed matter. His theory of Raman effect is a pioneering work in the field. Lev Landau and George Placzek derived the Landau-Placzek formula for the ratio of intensities for the Brillouin and Rayleigh scatterings of light. Hans Bethe and George Placzek provided a fundamental theory of neutron absorption resonances, deriving important laws and selection rules. Papers by Niels Bohr, Rudolf Peierls and George Placzek deal with the theory of nuclear reactions and rank among the world's classics. The well-known optical theorem bears the names of Bohr, Peierls

---

[8] A. Gottvald, "Kdo byl Georg Placzek (1905-1955)," *Čs. čas. fyz.*, Vol. 55, No. 3, 2005, pp. 275-287.

[9] Some additions absent in the Czech version belong to both authors.

[10] See http://dumbell.physics.muni.cz/placzek/papers/abstracts.pdf. Placzek was born in Brno on September 26, 1905.

and Placzek. In 1936, in a series of experiments, Otto Frisch and Placzek discovered that the absorption of neutrons in matter is strongly dependent on the atomic mass and the neutron velocity. For slow neutrons and light elements the neutron-capture cross section is inversely proportional to the velocity.

On the eve of World War II, Placzek's contribution was instrumental to the inception of the nuclear fission theory, an example of his many contributions still not properly appreciated.

After Lise Meitner's [11] forced escape to Sweden she started working with Otto Frisch on theoretical interpretation of the results of nuclear bombardment of heavy nuclei which had been obtained in Berlin by Hahn, Strassmann and herself. In his memoir *What Little I Remember* [12] Otto Frisch recollects:

> [Our] paper was composed by several long-distance telephone calls, Lise Meitner having returned to Stockholm in the meantime.[13] I asked an American biologist who was working with Hevesy what they call the process by which single cells divide in two; "fission," he said, so I used the term "nuclear fission" in that paper. Placzek was skeptical; couldn't I do some experiments to show the existence of those fast-moving fragments of the uranium nucleus? Oddly enough that thought hadn't occurred to me, but now I quickly set to work, and the experiment (which was really very easy) was done in two days,[14] and

---

[11] Lise Meitner was a part of the Hahn-Strassmann experiment in which various heavy elements were bombarded with neutrons. They observed that while the nuclei of most elements changed somewhat during neutron bombardment, the uranium nuclei changed greatly and broke into two roughly equal pieces. Lise Meitner, the Austrian citizen, was Jewish and could not stay in Germany after the 1938 *Anschluss* of Austria. Hahn and Strassmann stayed in Berlin and published a paper without mentioning Lise Meitner's participation. Lise Meitner and her nephew Otto Frisch were the first to explain the experimentally observed phenomenon as nuclear fission. In fact the very term "fission" was introduced by Frisch who borrowed it from biology.

[12] Otto Frisch, *What Little I Remember*, (Cambridge University Press, 1979), page 117.

[13] Otto Frisch was in Copenhagen in January 1939.

[14] In a letter to Lise Meitner dated January 17, 1939, Otto Frisch mentions that it took him three days to complete this experiment, see Roger Stewart's note in Sabine Lee, *Sir Rudolf Peierls: Selected Private And Scientific Correspondence*, (World Scientific, Singapore, 2007), Volume 2, page 878. In what follows, these two monumental Volumes are cited as "Sabine Lee." The above-mentioned experiment was carried out around January 13, 1939.

a short note about it was sent off to *Nature* together with the other note I had composed over the telephone with Lise Meitner.

In 1943-1946, Placzek occupied a leading position in the Manhattan Project, forming and later leading its Montreal Theory group at Chalk River. In 1948, he obtained a tenure at the Institute for Advanced Study in Princeton.

In this book we will try to make the next step in revealing so far unknown details of the adventurous life of George Placzek.

***

As was mentioned the only existing detailed biography of Georg Placzek was compiled by one of the authors and published in Czech more than a decade ago. Since then new documents pertinent to Placzek's life were found: in particular, a rather large number of letters written by him or sent to him by his friends and colleagues, referring to the pre-war years in Europe and post-war years in Los-Alamos and Princeton in the United States. In addition we managed to obtain some previously unpublished documents pertinent to Placzek's visits to the USSR and Palestine. In this respect, of paramount interest is the 1937 arrest and interrogation file of Konrad Weisselberg obtained by Alexander Kharlamov from the Ukrainian KGB Archive in 1996. It contains so far unknown facts about Placzek. So does the write-up of an interview with Alexander Akhiezer, who was a graduate student at the time of Placzek's visit to Kharkov. Using the above documents we managed to reconstruct, more accurately than before, Placzek's timeline (see page 61).

We managed to document the circumstances of Placzek's nine-month visit to Hebrew University of Jerusalem in 1935 (pages 148, 226, 231, 245) and the fact that in late November 1936 Placzek was received in Moscow by Nikolai Bukharin,[15] (pages 126, 132, and 164). It is not ruled out that the first encounter of Placzek with Bukharin may have occurred in 1933, during Placzek's first visit to Kharkov,

---

[15]Nikolai Bukharin (1888-1938) was a high-ranking Russian Bolshevik revolutionary, Soviet politician and prolific author on revolutionary theory. In 1924-34 Bukharin was a member of the ruling party Politbureau, and since 1932, became one of the key leaders of Heavy Industry. In 1934 Stalin appointed him as the Editor-in-Chief of the newspaper *Izvestia*, the second in importance in the USSR. On Stalin's order he was dismissed from his positions and arrested by the NKVD on February 27, 1937, and then sentenced to death by a staged "public" trial. He was executed by firing squad on March 15, 1938.

through Alexander Weissberg who was in charge of construction of the UPTI[16] Cryogenic Laboratory and in this capacity was in contact with Bukharin on a regular basis. In Placzek's letter of September 4, 1938, (page 164) it is mentioned that he had "solicited for Landau, trying to convince Bukharin that they should now and then let him travel abroad."

It became possible to established Placzek's chronology during the war years much more accurately than it was known previously. On page 92 the reader will find Placzek's letter to Charlotte Houtermans dated January 28, 1939. This letter was written on board the *S. S. Normandie* when Placzek was on his way to the United States. Thus, his tenure at Cornell University started in early February 1939. In the letter to Hans Bethe of July 31, 1942, (page 170) Placzek informed Bethe that he was about to leave Ithaca for the United Kingdom. From the Los Alamos document of June 22, 1945 (page 237) we infer that from August till November 1942 Placek worked at the Cavendish Laboratory, Cambridge University, UK, on chain reaction problems. Finally by December 1942 he was already in New York, interviewing candidates for the Theory Division of the Montreal Laboratory. Bengt Carlson recollects: "I had been interviewed by Placzek just before Christmas, 1942, in a hotel near the Grand Central Station in New York City."[17] Placzek was among the first to arrive at Montreal Laboratory in Late December 1942, where he continued conducting interviews in a bid to hire as strong and motivated theorists as was humanly possible.[18]

In the Shelby White and Leon Levy Archives Center (Institute for Advanced Study, Princeton, NJ, USA) we found a letter dated March 30, 1955, by Stefan Rozental (who at that time was the Head of the Theoretical Study Group of CERN, with the Headquarters still in Copenhagen rather than Geneva), presenting a direct proof of high visibility of George Placzek in the physics community in the years preceding his death (see page 220).

---

[16]Ukrainian Physical Technical Institute in Kharkov.

[17]See M.M.R. Williams, *The Development of Nuclear Reactor Theory in the Montreal Laboratory of the National Research Council of Canada (Division of Atomic Energy)*, Progress in Nuclear Energy, Vol. 36, No. 3, pp. 239-322, 2000, page 8.

[18]M.M.R. Williams in his report, see the previous footnote, quotes George C. Laurence: "The first of the staff from England arrived about the end of the year 1942. They were P. Auger and B. Goldschmidt of France, *G. Placzek* of Czechoslovakia, S.G. Bauer of Switzerland, H. Paneth and H.H. Halban of Austria and R.E. Newell and F.R. Jackson of Great Britain. We temporarily occupied an old residence at 3470 Simpson street belonging to McGill University.

In the existing literature the circumstances surrounding the last days and the death of George Placzek in Zurich in 1955 are foggy, to say the least. Various sources presented in Chapter 5 either keep silent or give rather incomplete and inconsistent information.

We located a letter from Wolfgang Pauli to Erwin Panofsky [19] dated October 12, 1955, (i.e. three days after Placzek's death) which contains some first-hand evidence of this event. More importantly, the 1955 Zurich Police Report $\mathcal{N}^{\underline{0}}$ 16420/55 regarding the circumstances of Placzek's death became available (pages 342-343) providing us with exhaustive information which the reader will find below in this book.

$$***$$

General Acknowledgments

We would like to thank [20] the following archives and individuals for providing us with documents and recollections regarding Placzek's life and particular events. We are grateful for permission to use these materials in the present Volume:

Houtermans Family Archive (Giovanna Fjelstad, Jan Houtermans, Annika Fjelstad);

Eva Zeisel Archive (Jean Richards, Judith Szapor);

Ernest Kopp (Zürich Police Report);

Karl von Meyenn (Pauli's letter);

Yuri Ranyuk (materials related to Kharkov, including Akhiezer's interview);

Art Kharlamov (Weisselberg's file);

The Peierls Family Archive (Gaby Gross, Jo Hookway, Tim Peierls, Sabine Lee);

Emilio Segrè Visual Archive of the American Institute of Physics;

Niels Bohr Archive, Copenhagen (Felicity Pors and Rob Sunderland);

Heisenberg Family Archive;

---

[19] Courtesy of Prof. Ernest Kopp; see page 54.
[20] See also page 59.

Cécile DeWitt-Morette Archive;

American Physical Society;

American Institute of Physics Niels Bohr Library (Audrey Lengel);

Bradbury Science Museum – Los Alamos National Laboratory (Alan Carr);

The Central Archive of Hebrew University of Jerusalem (Michael Vinegrad);

The NRC Archives, Canada (Steven Leclair);

The National Archives (TNA), Kew, UK;

Bodleian Library Archive, Oxford University;

Shelby White and Leon Levy Archives Center, Institute for Advanced Study, Princeton, NJ, USA (Casey Westerman and Erica Mosner);

RGASPI, Moscow, Russia;

Sabine Lee (Placzek-Peierls and Bethe-Peierls correspondence);

Jan Fischer (Prague);

Chris DeWitt (Austin, TX);

Issachar Unna (Jerusalem);

Ella Andriesse (New York);

Gennady Gorelik (Boston, MA).

*****

We highly appreciate invaluable work on translating certain documents from German, French, Italian, and Russian carried out by Vladimir von Schlippe, Alexander Tschernow, Lyubov Chernova, Daria Monakhova, Alice West, Lena Funcke, Valentina Maslova, Sean Kalafut, Maxim Konyushikhin, Stefano Bolognesi and Marco Peloso. It is impossible to overestimate Sabine Lee's assistance, including multiple consultations and correspondence with the authors, and her general encouragement and generosity. Thank you, Sabine.

We express our sincere gratitude to our World Scientific Editor, Lakshmi Narayanan, and to Janice Sim for their invaluable assistance and patience.

Graphic design of this book is by Polina Tylevich. The authors are grateful for her excellent and insightful work.

Contents

List of photographs and graphic material

Fig. 5.2, page 256: Placzek's photograph from his Guggenheim Foundation application file, 1954 or 1955

Fig. 5.3, page 275: Sketch of Placzek by Otto Frisch

Fig. 5.4, page 284: Bradbury Museum of Science in Los Alamos

Fig. 5.5, page 295: Eva Striker in the USSR, 1934

Pages 298 and Fig. 5.6 on page 299: Van Hove's Oration title page and Placzek's picture from the Oration

Fig. 6.1, page 321: Journal of Jocular Physics (October 7, 1935), title page

Fig. 6.2, page 323: 1935 Niels Bohr Festschrift. Table of Contents

Fig. 6.3, page 329: Placzek's last visit to Berlin in 1936 (tentatively)

Pages 330 and 331: Wolfgang Pauli and a diagram of forces, 1931

Fig. A.1, page 338: Els as a teenager

Fig. A.2, page 339: Els with her grandfather Samuel van den Bergh

Fig. A.3, page 341: Els Placzek, February 20, 2005

Fig. A.4, page 342: The first page of the Zurich Police Report of January 25, 1956

Fig. A.5, page 348: The title page of Placzek's "Raman Effect" in Russian, Kharkov, 1935

Fig. A.6, page 349: Theory Division at Los Alamos, May 10, 1945

Figs. A.7 and A.8, pages 350, 351: N. Bohr, G. Peierls, and G. Placzek, *Nature*, Vol. 144, $N^{\underline{o}}$ 3639, 200 (1939)

Pages 352-363: G. Placzek, "The Scattering of Neutrons by Systems of Heavy Nuclei"

PART 1

GEORG GEORGES GEORGE PLACZEK: LIFE AND FATE

The Story of Georg Placzek (1905-1955)

Aleš Gottvald

Introduction

"He was one of the most remarkable people that I have met in my life"
– as Victor Weisskopf recollects [1; 2]. "Institute's always stimulating
Czech" – as Niels Bohr remembers him [3; 4]. "Bohemian in every
sense of the word" – as Otto Frisch says with a smile [5; 4]. No, we
should not begin from quoting a long and almost incredible gallery
of warm superlatives that many of the distinguished theoretical and
experimental physicists of the 20th century used when recalling Georg
Placzek [6; 7; 8; 9; 10; 11; 12; 13]. Anybody who encounters the name of
theoretical physicist Georg Placzek for the first time might easily think
that somebody like him could hardly be real, otherwise we would had
heard about him before. In the fate of Georg Placzek, in a nutshell,
one can find both the brightest highlights and deepest downfalls of
the history of the past century. Very likely, Georg himself would not
feel comfortable with too much attention and celebration devoted to
him. We imagine him sitting modestly in some of the remote seats in
the busy lecture room of Niels Bohr's institute in Copenhagen, letting
his colleagues and friends – the most famous physicists of that time –
occupy the first row. And he offers us an opportunity to remember him
with awe and admiration on the occasion of his two life anniversaries in
the World Year of Physics: 100 years from his birth and 50 years from
his passing away...[1]

Who was the physicist Georg Placzek in a nutshell? A creator of
many fundamental theories: of the Raman scattering, spectra of polyatomic molecules and the light scattering in liquids and gases, diffusion
and moderation of neutrons, and inelastic scattering of neutrons on
crystals... Co-author of the first chapters of neutron physics. Co-author
of a direct experimental proof of nuclear fission. Catalyst of landmark
ideas leading to the first atomic bombs and nuclear reactors. An inspiring and capable leader of the Montreal theoretical group involved
in the Manhattan project during World War II.

His work in any of these domains would secure him a lasting place in

---

[1]The core of this section was written on the occasion of the Symposium in Memory of Georg
Placzek held in Brno (Czech Republic), September 21-24, 2005. A number of revisions and additions were made in collaboration with M. Shifman in 2017.

a scientific gallery of honor. However, the true contribution and importance of Georg Placzek does not come just from his list of publications (see [2]). For many of his colleagues, he was a much-sought-for collaborator, critic, adviser and co-author – even for some most distinguished physicists of his time, who later became Nobel prize winners. He was a kind of scientific "consciousness" for his colleagues whose many works he stimulated. He was a Renaissance personality with a large scientific and human perspective. He was never boring and lives in tales of physics as a generator of many funny "Placzek Geschichten."[2] He was an active witness of many historical events. Let us take a walk through his life...

The landscape of Georg's childhood: Brno, Alexovice

Georg Placzek was born on September 26, 1905, at the very center of Brno,[3] in a house at Grosser Platz №3 (today *Náměstí Svobody*). He was the oldest of three children in a large, prominent, and wealthy Jewish family. In 1904, his father Alfred Placzek (1870-1942) became a co-owner (*komandista* and *prokurista*) of an internationally renowned textile factory "Skene a Co." in Alexovice near Ivančice. The factory had been established around 1845 by the Baron Alfred Skene (in the years 1864-66 a mayor of Brno). In a Memorial book of Alexovice [14] one can find many moving remarks about Placzek's family, which leaves no doubt that the family changed the fate of the village and whole region in a very remarkable and praiseworthy way. Thanks to Alfred Placzek, the harsh Germanization of Alexovice, supported by the Skene family, was stopped. In 1913, Alfred Placzek took over the factory from his partner Louis Skene completely.

During World War I, the Placzek's textile factory obtained significant state contracts, which saved many people in Alexovice from the war and death. The factory was prosperous and Placzek's family strove to use their wealth for the benefit of Alexovice citizens during the difficult war times. When starvation hit Alexovice by the end of the war, Alfred

Figure 1.1   Bohemia and Moravia.

____

[2] Placzek stories.

[3] At the time of Placzek's birth this city was known as Brünn, in Moravia.

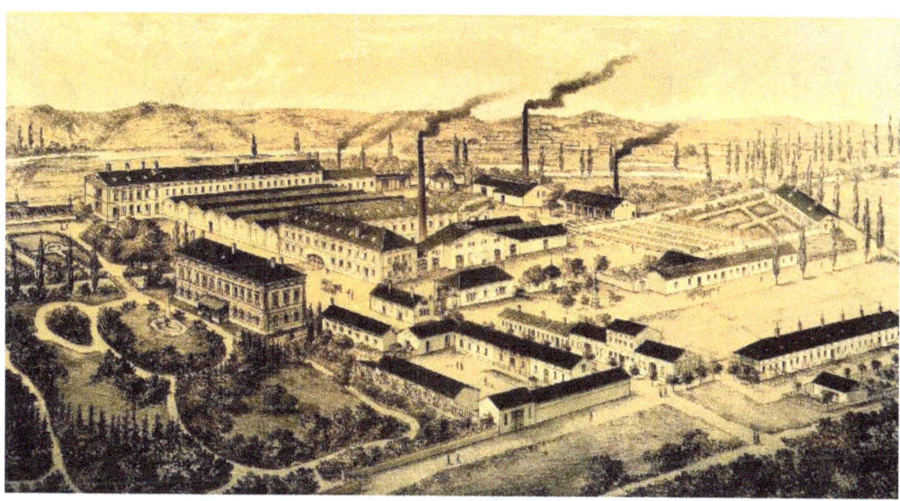

Figure 1.2   A view of the Skene & Co. textile factory in Alexovice, 1879. On the left, in front of the park area, is the house of baron Alfred Skene, later inhabited by the Placzeks (after 1913).

Placzek devoted "unlimited financial resources" to buy some food for the starving. When even money became of no use, every week some cattle were butchered at the site of the Placzek factory and the meat was sold to the employees for a low price. Georg's mother Marianne (née Pollack) (1882-1944) knew by name all women working in their factory and their children, and she personally handed out some food and gifts. She was an optimistic and smiling woman, interested in theatre, antiques, and traveling. In December 1922, Alfred and Marianne Placzek obtained honorary citizenship of Alexovice, "for their philanthropy, which they showed to indigent people." In April 1938, Alfred Placzek was also named a honorary member of a regional youth care association in Ivančice, "as a long-time supporter and sponsor of its human efforts."

The Placzek family was summering in Alexovice and wintering in Brno. The garden of the house №4 in Alexovice,[4] just in front of their factory, witnessed the childhood games of Georg and his one-year younger brother Fritz. In 1913, the family moved to a large beautiful house in a park adjacent to their factory, which had been used by the Skene family before them (Fig. 1.2). Both houses can still be found in Alexovice. The Placzek family was well integrated to a mixed Czech-German-Jewish environment. Though their first language was German,

---

[4]Currently, Tovární 4/14, Ivančice-Alexovice.

Figure 1.3   The house at Alexovice, where the Placzek family lived from 1913 to the beginning of the German occupation in March 1939. Sitting on the bench on the left is Georg's sister Edith. Courtesy of J. Kocourková and F. A. Placzek.

their contemporaries recall young Georg speaking in Czech with girls in the kitchen. Young Placzek's children were supposedly educated by a French tutoress [15].

Inspiration to science: Dr. Baruch Placzek[5]

Who from the closest family might have inspired young Georg towards his future scientific career? It had to be a very strong inspiration, taking into account that the first-born son was likely expected to take over his father's factory.

An extraordinary personality that, as we suspect, put Georg's life on a scientific trajectory, indeed was a close relative. It was his paternal grandfather, Dr. Baruch Placzek (1834-1922), a man renowned for his wisdom, education and Renaissance interests. For many years, he served as the Rabbi of Brno (since 1860) and later the Rabbi of Moravia

---

[5]Most of information about Baruch Placzek in this section comes from Dr. Ruth Davis, who specializes on his biography and the Jewish history of Brno.

Figure 1.4   Dr. Baruch Placzek, the Rabbi of Moravia, Georg's grandfather and his inspiration to the scientific way of life. Credit: Archive of M. Fuhrmann.

(from 1884 to his death). In his secret private life, however, he was an enthusiastic amateur scientist. He was in correspondence with Charles Darwin regarding evolutionary theory. He was also a close friend of Johann Gregor Mendel, with whom he used to play cards and, according to some family legends, perhaps was the only person of his time who recognized the importance of Mendel's results in heredity (genetics). After Mendel's death, Baruch Placzek might even have continued with some of Mendel's experiments – but this information (coming from Georg's eldest aunt Sarah) is uncertain. With much more written evidence we can say that Baruch Placzek was an environmentalist and a pioneer of ecology. He published many works, ranging from poetry and prose (under a pseudonym Benno Planek) through impressive obituaries of some prominent Jewish citizens of Brno, to his scientific works (mostly from ornithology). Unfortunately, probably lost forever are his scientific diaries, which might also clarify his links to Mendel: their trace was lost during World War II in Alexovice [16] ...

High School years

We see twelve-year-old Georg for the first time in a rare family photograph from April 1918 (Fig. 1.5). The date above can be determined

Figure 1.5   The Placzek family in April 1918. Standing in the middle is Georg Placzek who is twelve years old. In the first row (from left to right) standing is his brother Fritz, his mother Marianne with his newborn sister Edith, and the rightmost is his father Alfred Placzek. Courtesy of J. Kocourková and F. Anthony Placzek.

because – together with his father Alfred, his mother Marianne and his younger brother Fritz – his newborn sister Edith is also pictured. It is the only known photograph of Georg, which we have from his "Brno-Alexovice era." Young Georg is shown here several months before entering (in September 1918) the Deutsche Staats Gymnasium in Brno, nicknamed also the "seminary of science." It is today the building of JAMU (Janáček's Academy of Music Arts) on Komenského nám. No. 6. This Gymnasium Georg attended from "tercie" (1918-19) to "oktáva" (1923-24).

From class files, which survived in Brno archives, one can learn many interesting details. First – Georg's study results: in German, Greek and Latin, Georg is mostly classified by the best grade (excellent), whilst in physics and mathematics – he is mostly "good." Every year he is taking Czech language as an option, and his grades are "good." Though he finished his high school with honors, some of his school mates showed better results. We smile reading a remark in his "oktáva" files, which suggests that he was "opposing with something undue" to some of his professors. Many later testimonies agree about Georg's "sharp satirical tongue." By the way – even his early career as a scientist was probably "pushed through" in his family using his eloquent tongue, and this strong verbal capability would follow him all his life.

The high school files also include addresses which Georg (at least formally) used as a student. There is surprisingly many of them, almost anticipating Georg's extraordinary globetrotting in his later years, incomparable to most of his colleagues. At that time, the major house of the Placzek family was situated at the address Talgasse 20 (currently, Údolní street). However, Georg listed also other short-time addresses: first, at Anastasius Grün Gasse 5 (today Hilleho street), then at Eichhorngasse 18 (today Veveří street). But he also shows his father's address "Alexowitz bei Eibenschitz" (Alexovice u Ivančic). From old phone books of Brno we also learn of several other addresses linked to the Placzek family. Georg's grand-father Baruch Placzek occupied a house at Talgasse 7, and later (from 1911) a house at the address Tivoligasse 38 (today Jiráskova street), together with the family of Georg's youngest aunt Irma Seidler. The family of his oldest aunt Sarah Türkl used a house at Talgasse 22. An administrative building of the Placzek factory "Skenc & Co." was situated at the address Cejl 5. Alfred Placzek also owned a building at Orlí 28. All of these are large brownstone houses, showing the wealth of their inhabitants. Let us note that "our" Placzek family is not closely related with another important family of the Placzeks in Brno, who owned at that time a large textile and clothing house "Plaček" on Masarykova street, well known to the Brno citizens.

Probably around 1925, a large villa at Kounicova 18 became the major living house of the Placzek family in Brno. This villa stood just in the neighborhood of today's Janáček museum. Alfréd Placzek and his family used this villa until the tragic days of the year 1939, when it was "confiscated for the Deutsches Reich." Ironically, under the German occupation it became the Gestapo Headquarters in Brno. This building was destroyed by the US Air Forces on November 20, 1944. Some rare photographs of this villa, probably taken before 1929, miraculously survived (Fig. 1.6). This and some other rare photographs, together with some invaluable testimonies, were preserved by Ing. arch.[6] Jarmila

---

[6]Inženýr architekt, Engineer architect (from Latin ingerum architectus).

Figure 1.6   Placzek's villa in Brno, Kounicova 18, circa 1930. During the war tentatively it was used as the Brno Branch Gestapo Headquarters. The house was destroyed at the end of WWII. Credit: Archive of J. Kocourková.

Kocourková, whose father served as a private driver for the Placzek family. Besides many other things, these historical photos show a whole "fleet" of beautiful cars, another evidence of high economical standing of the family.

We have several testimonies that Georg Placzek was attending his family in the years before World War II, in addition to working and traveling abroad most of his time. Ms. Dorothea (Dorrit) Forman (née Fuhrmann; born in 1917, a daughter of one of Georg's cousins), recalls her meeting with him in Alexovice. Georg came from his travels, already an experienced (but still not famous) physicist. Ms. Forman remembers how deeply Georg impressed her during that meeting: "A true personality, there can be no question about that." This testimony, though pointing to a meeting rather uncertain in time, is worth mentioning. Practically all future testimonies of Georg's friends and collaborators unanimously agree that his personality impressed them strongly and deeply.

Ms. Kocourková, the Placzek's personal driver's daughter, remembers a meeting with Georg in Alexovice. A nice tale about this a meeting later lived in her family: Georg came to Alexovice from England, wearing a very fancy wool jacket. His father Alfred tested the outfit with his fingers and Georg commented on it: "What a material, a true English kamka!" "Yes," his father replied, "but made here in Alexovice!"

Other testimonies from the circle of his family and contemporaries show that young Georg advocated radical left-wing opinions and criticized a lifestyle of "bourgeoisie." At this point, he evidently resonated very well with his good friend, young Lev Landau, whose "revolutionary communist" ideology is well documented from his stays with Bohr in Copenhagen to his politically forced escape from Kharkov. No wonder that young Georg was treated as a "dissident" in his rich industrial family. Both Placzek and Landau were "cured" from their "revolutionary" ideology only after their personal harsh experience with Stalin's purges and Soviet reality. This first-hand experience of Placzek with the iron-fisted communist regime of Stalin will show its important historical consequences soon and influence attitudes and actions of some key persons in the Manhattan project (especially Oppenheimer and Teller).

Here is another illustration to the life of the Placzeks before the Second World War, as narrated by Ms. Kocourková and confirmed by some written evidence. Alfred Placzek and his family behaved in a very familiar and caring way to their personal driver's family and other loyal employees. Hence, any stereotypical image about "cruel capitalists" is far from reality in their case. It is a tragic paradox that several German employees of Placzek's factory in Alexovice repaid them very badly after Hitler came to Czechoslovakia [16]...

University studies

Let us return back to the time in Georg Placzek's life after his high school study. We have rather sketchy information about his university studies. From September 1924 to July 1928 (for 5 semesters) he was studying at the Vienna University, with an interruption of three semesters in the middle which he spent at the German University in Prague (the reasons for this "switch" are unknown). His dissertation work (rather diploma work from the contemporary perspective), entitled *Zur Dichten – und Gestaltbestimmung submikroskopischer Probekörper (Versuche im inhomogenen elektrischen Feld)* was elaborated under the guidance of Prof. Felix Ehrenhaft, at the Third Institute of Physics of Vienna University. Here Placzek defended it with honors in July 1928. This dissertation work was focused on the topic of electrostatics in matter, and was mostly experimental. Later in life Georg was joking about this connection: "Actually I started as an experimental physicist, but later I stopped working and thus some people think that I am a theoretician." In Vienna, Georg also passed two doctoral exams (the so-called "rigorosum" and "philosophicum"), both with honors.

And this begins Georg's monumental pilgrimage between numerous prominent centers of physics of that time. In those years, many famous and distinguished physicists became his close friends and collaborators: Werner Heisenberg, Enrico Fermi, Edoardo Amaldi, Emilio Segrè, Hans Bethe, Edward Teller, Rudolf Peierls, Felix Bloch, Lev Landau, Niels Bohr, Victor Weisskopf, Otto Frisch, Subrahmanyan Chandrasekhar, George Gamow, John Wheeler, Robert Oppenheimer, Léon Van Hove,... His native and familiar name "Georg," which we utilize in this essay, depending on the country of his residence has been changed to "Georges" or "George" – by which he is best known in the world.

Utrecht, Leipzig, Göttingen, Rome

Young post-doctoral associate, Georg Placzek, began his career as a physicist with a two-year stay around Hendrik Kramers and Leonard Ornstein in Utrecht, the Netherlands. Judging from his first papers in scientific journals, it is here that he starts elaborating his theory of Raman's scattering, his major topic for some five years to come. The theory soon became an appreciated and highly cited contribution to physics. Very likely, his penetrating choice of this topic was stimulated by a series of short experimental papers by Raman, issued from April to the end of 1928 in *Nature*. This physicist from India was to be honored for these discoveries by the Nobel Prize very soon – in 1930. In contrast, Placzek never obtained any formal appreciation for his excellent underlying theory. Some historians of science show this case

"Raman versus Placzek" as a demonstration of the fact that theorists are rather neglected in the "Nobelization" procedures, in comparison with their experimentalist counterparts.

In Utrecht, Placzek also published his two older works from electrostatics, one of them being his dissertation from Vienna. The stay in Utrecht opened doors to Niels Bohr in Copenhagen (Kramers was an indispensable "first assistant" to Bohr for many years, and Placzek would obtain a similar position in Bohr's Institute several years later). His stay in Utrecht also resulted in an offer from the Hebrew University (in this matter, Ornstein took some role in Placzek's nomination).

In those years, quantum mechanics was still a newly discovered exotic butterfly that lured many into new applications with its tempting wings. The youngest German professor, Werner Heisenberg from Leipzig University, is one of the leading quantum "magicians" and Georg Placzek appears in his Lab around the end of 1930. He helped Heisenberg with his regular Thursday seminars on physics. The stay is documented by two rare photographs. One of them (Fig. 1.7) features Placzek, Heisenberg, Peierls, Bloch, Gentile, Wick, and Sauter. We can determine its date quite reliably – pay attention to a black ribbon on Heisenberg's arm, to mourn his father's death (on November 22, 1930). There is no need to emphasize what prominence these people were soon to achieve. One psychological detail is worth mentioning: the whole group (but Placzek) is looking to the camera with obvious self-confidence, whilst Placzek is looking somewhere apart with some nostalgia... "He was somewhat different ..." recalls Gilberto Bernardini [17]. "He was extraordinarily good and brilliant as a physicist, with surprisingly extensive cultural and intellectual interests."

In the second photograph (see page 18) Heisenberg and Placzek play ping-pong together, whilst Bloch is watching. By the way, Heisenberg was very enthusiastic about the then-new sport of table tennis.

Georg Placzek entered those new applications of quantum mechanics (later somewhat sarcastically called "Knabenfyzik"[7]). He created a beautiful theory of Raman's scattering of light on polyatomic molecules. It addressed an explanation of combinatorial frequencies in the spectra structure, that are shifted from the central Rayleigh line.

This work by Placzek elucidates a fundamental role of symmetries and group theory for describing the spectral phenomena. This group-theoretic line of reasoning would later be developed by Eugene Wigner

---

[7]Boyish physics, or physics made by boys.

Figure 1.7 Heisenberg's visiting fellows at Leipzig University at the end of 1930. Left to right in the front row are Peierls and Heisenberg, standing behind them are Gentile, Placzek, Wick, Bloch, Weisskopf, and Sauter. Credit: http://www.heisenbergfamily.org

and Laszlo Tisza (to whom some inspiration from Placzek was conveyed by Edward Teller). Somewhat later (in 1933), Placzek extended this work towards statistical thermodynamics, together with Lev Landau (whom he probably got acquainted with while in Germany).

Amongst all those friendships with young talented physicists, the one with Edward Teller played a particularly important or even fatal role for both of them. Their paths intertwined in several periods of their lives, with deep impact on Teller and Placzek. At the beginning of their career, Placzek strongly influenced Teller's turn to nuclear physics, introducing him to Fermi in Rome. Let us quote Teller's own recollections verbatim:[8]

> Placzek wanted to continue with his work in the holidays while Göttingen was closed. I wanted to go home to Budapest and he said: No. I, Placzek, want to visit

---

[8]From Edward Teller's recollections *Going to Rome with Placzek to visit Fermi*, The Web of Stories, http://www.peoplesarchive.com

with Fermi in Rome and you come along. In a way, I
was interested but I had just started to make my own
money, needed little help from home. Didn't quite know
how to pay for my stay in Rome. Oh, said Placzek – I
will take care of that. I'll ask Fermi. Here I get a copy
of a letter from Fermi, that he has written to appropriate authorities in Hungary – I hear that Doctor Teller is
considering to visit Rome for a few weeks. He's a very
famous physicist, and I want his cooperation. Could you
please help him to get to Rome and to stay there? Fermi
and I never met. That he had reason to consider me as a
famous physicist was, to say the very least, an impudent
exaggeration. But what makes the story particularly en-
joyable for me is that together with the copy of the letter
to the Hungarian authorities, I got, attached, a little note
from Fermi: "Dear Doctor Teller, I am sending you this
copy. I want you to know that actually I would be really very happy to see you in Rome." So he took back
the exaggeration but replaced it with a premature offer of
friendship which, of course, became a very real friendship
in the course of time.

Placzek nicknamed Teller a "molecular inspector." During their stay
in Rome in 1932, he "educated" him by some merciless jokes, which
did change Teller's behavior (as Hans Bethe recalls). But their close
relationship was obviously not hampered, rather the opposite. Three
years later, Placzek nominated Teller as his assistant at the Hebrew
University in Jerusalem – but this was never realized (see below). It
was also Placzek, together with Weisskopf and Tisza, from whom Teller
obtained first-hand information about the Stalinist Soviet regime – a
fact which was decisive for his future attitudes and actions in the Manhattan project and later H-bomb development in the time of the Cold
War. Finally, not without reason we can suppose that broken personal
relationships in the triangle Placzek-Teller-Oppeheimer were fatal for
the final period of Placzek's life...

But let us return to Rome of 1931-32, to a group of young physicists
around Enrico Fermi. Together with Teller, Placzek's collaborators
and close friends here were Hans Bethe, Edoardo Amaldi, and Emilio
Segrè. As recollected by Gell-Mann,[9] Placzek was nicknamed Il Santo

---

[9] In: *Fermi Remembered*, Ed. James W. Cronin, (The Univ. of Chicago Press, Chicago-London,

(and Amaldi was called Il Cardinale, and Fermi was called Il Papa). An interesting testimony was preserved by Franco Rasetti [8]. It addresses the fine task of writing and publishing scientific results, a particularly tough problem which Georg Placzek faced his whole life.

Figure 1.8 Georg Placzek and Werner Heisenberg playing table tennis, with Felix Bloch standing next to the door and watching them. Leipzig University, late 1930. Credit: http://www.heisenbergfamily.org

Placzek and Bethe were sitting side-by-side at their writing desks. Placzek was writing his fundamental extensive work on Raman effect for the "Handbuch der Radiologie" [18], while Bethe was writing his famous paper on one- and two-electron systems for "Handbuch der Physik." The way Bethe was able to write his papers was extraordinary: sitting stiff as a board, he wrote his text without interruption, without scratching or changing any single sentence. Not so Placzek: he was very unhappy about everything he wrote, struggling with every formulation and rewriting his text many times. "He was crazy from the

2004), see Ch. 6 by Murray Gell-Mann.

fact that Bethe is able to write his work without interruption, whilst he had to rewrite every page ten times before he was satisfied with it." [8] In his struggle with publications, Placzek was very similar to Bohr or Landau, to mention just few famous physicists. However, Placzek had mostly no permanent assistant for writing or mathematics – such as Kramers for Bohr, or Lifshitz for Landau. In his later years, Placzek's troubles with writing would be further complicated by his escalating health problems (some sources indicate high blood pressure and heart disease, but – most likely – a bipolar mental disorder was his gravest problem, see Appendix, pages 343 and 344).

While writing scientific papers was never easy for Placzek, his capability concerning foreign languages was enormous and generally admired. A strong background for linguistic skills was already established during his high school studies in culturally-mixed Brno. But Placzek was extending his command of foreign languages gradually, absorbing fluently the languages of the countries where he lived. As a result, he eventually spoke, more-or-less fluently, about ten foreign languages, including many "dry" wise cracks. In particular, he mastered Italian "just like that" by reading Decameron [9].

Copenhagen, 1932-38

Niels Bohr's Institute in Copenhagen became a second home of Georg Placzek in the years before the Second World War. After withdrawal of his long-time assistants (Hendrik Kramers and later Oskar Klein), Niels Bohr was in need of somebody else, somebody with whom he might discuss his ideas at the blackboard, who might do some math for him, and possibly write "his" papers for him. For the first two of these three requirements he discovered an almost ideal person in his "always stimulating Czech." On the other hand, Placzek obtained a safe and inspiring asylum in the famous Niels Bohr Institute, a Mecca of modern theoretical physics in those times.

In the Bohr Institute, Placzek got acquainted with practically all of the distinguished theoretical physicists of his time. In a group of some fifteen international fellows, an "inner circle" of closer friends emerged. Together with Placzek, this "inner circle" was formed by Victor Weisskopf, Felix Bloch, George Gamow, F. J. Williams, Subrahmanyan Chandrasekhar, and Homi Bhabha [1]. His stay in Bohr's neighborhood is documented by some dozen photographs in Bohr's archive. One can recognize him on some photographs from famous "Bohrfests" in Copen-

Figure 1.9   Left to right: Paul Ehrenfest, George Placzek and Franco Rasetti in Rome, 1931. Photograph by Samuel Goudsmit, courtesy of AIP Emilio Segrè Visual Archives, Goudsmit Collection.

hagen – dated 1933, 1934, 1936, 1937,[10] (and also 1947), and also on a photo taken on the occasion of Bohr's 50th anniversary – in October 1935. On the photograph from the 1933 Bohrfest, Paul Ehrenfest sits

---

[10]See pages 22 and 48, respectively.

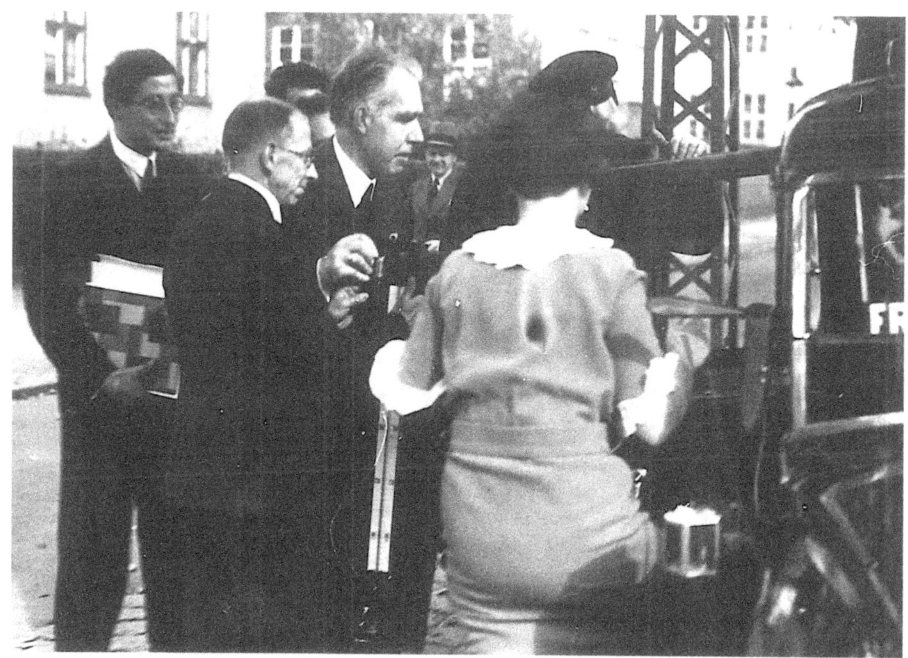

Figure 1.10   George Placzek, Holger Olsen, Niels Bohr, and Hilde Levi. Niels Bohr's 50th birthday celebration, 1935. Credit: Niels Bohr Archive, Copenhagen.

in the first row; he committed suicide soon after (September 25, 1933, Amsterdam).

Placzek's long stay in Copenhagen coincides with a period when major attention at Bohr's Institute gradually changed from the principles of quantum mechanics to nuclear physics – and newly discovered neutrons in particular. At that time, Placzek was again involved in experimental physics, cooperating with Otto Frisch. This cooperation was quite natural for both of them, as Placzek knew well what to measure, and Frisch was equipped with the instrumentation required [5; 19]. Moreover, Frisch was a nephew of Lise Meitner, who conveyed to him some private information about nuclear experiments of Hahn's group in Berlin.[11] Somewhat later, during Christmas 1938, such a flow of information between Meitner and Frisch would lead to a rationale for nuclear fission – with Placzek suggesting a direct experimental proof of the fission to Frisch [19; 10].

---

[11]It is known that Placzek visited Berlin even after Hitler came to power (certainly, in summer 1934 and probably also in 1936).

Figure 1.11 The 1937 Copenhagen Conference organized by Niels Bohr. Left to right: (first row) N. Bohr, W. Heisenberg, W. Pauli, O. Stern, L. Meitner, R. Ladenburg, J. Jacobsen; (second row) V. Weisskopf, C. Møller, H. Euler, R. Peierls, F. Hund, M. Goldhaber, W. Heitler, E. Segrè, S. Hoffer-Jensen; (third row) G. Placzek, C. Weizsacker, H. Kopfermann, [...]. Credit: Photograph by Nordisk Pressefoto, Niels Bohr Institute, courtesy AIP Emilio Segrè Visual Archives, Fermi Film Collection, and Niels Bohr Archive, Copenhagen.

Figure 1.12   Georg Placzek, 1938. Credit: Niels Bohr Archive, Copenhagen.

In 1936, Placzek and Frisch published the article "Capture of Slow Neutrons" in Nature. They measured an absorption of slow neutrons in cadmium, boron and gold (for the golden targets Placzek suggested using the Nobel medals that some Jewish physicists – fleeing from Hitler's Germany – deposited with Bohr; these medals were made of high quality gold). In their paper, Placzek and Frisch reported that the resonant levels of the neutron absorption show up at surprisingly low energies.

Placzek was impressing his friends and colleagues with his capacity to discuss a broad range of questions, ranging from physics and culture to polity. He had a sense for the tasks of the time, often recognizing the substance of some problem earlier than the others. However, his private life he could not manage so well. He was a night owl, often working or carrying on Frisch's experiments till 3 am. Sometimes around noon, he would appear in his crumpled pajamas in the Institute's library, a funny look to Danish girl-students and not-so-funny look to Mrs. Bohr. To silence wicked tongues, we observe that a good suit and a tie is much more typical for Placzek on practically all historical photographs. Fond memories about those Copenhagen years of Georg Placzek belong to one of his best friends – Victor Weisskopf [1], and also to Otto Frisch [5]. We shall not to steal their "Placzek Geschichten" – but one from Weisskopf:

"One year in Copenhagen, Bohr's international fellows prepared a funny fable where some roles of animals were created for the individual famous physicists. For instance, Pauli played the role of the hippopotamus, Heisenberg was given the role of an antelope, and so on. But they faced a problem: they could not invent an adequate animal for Placzek. Finally, they invented a new animal, a new species simply called 'the Placzek.' Well, Placzek was unique..."

Kharkov, 1933

The first stay of Placzek in Landau's circle in Kharkov lasted for 7 or 8 months, from May to December 1933 (see Document 1.25 on page 67 and its transcript on page 234). Placzek probably met Landau in Germany some three years earlier, or possibly later in Copenhagen, during Landau's two short trips to Bohr's Institute in 1933. This was the very end of Landau's travelings abroad, in part due to Gamow's escape from the USSR in October 1933.

Physicist Alex Weissberg-Cybulski, an emigrant from Austria and an ardent communist at that time, was instrumental in arranging stays of foreign theorists in Landau's group.[12] There is some evidence that Placzek himself took care of the financial aspects (at least, in part) of stays for some foreign guests in Landau's circle. Georg's father, Alfred Placzek, reputedly joked about it: "I am sponsoring splitting of the atom."

Placzek, similar to Tisza, was impressed by Landau's general concept of theoretical physics. Thermodynamics was not just an old-fashioned domain of theory for Landau, but rather a vital and universal component which should be integrated to all theoretical physics. At that time, Landau and his disciples initiated their plan to develop a *Course of Theoretical Physics*, adopting this viewpoint. Placzek already had a deep and extensive theory of Raman's scattering developed. It was no wonder that the fruits of their collaboration was an extension of this theory towards thermodynamics and statistical physics. Formally, just a one-page article published in 1934, see Fig. 1.26 on page 68, originated from this joint work [21]. However, we find an elegant (though just verbally described) formula in it, later well-known in physics as the Landau-Placzek ratio. A fine structure of a spectrum of the scattered light is related to some thermodynamic quantities and entropy fluctuations in liquids and gases.

There is a note in the article that a detailed extension of the work should follow. As one can see in [21], such an extension never materialized (Placzek's classical work [29], issued in 1934, had obviously been written a year or two before in Rome). More fruits of this first collaboration between Landau and Placzek appeared only much later, in the celebrated *Course of Theoretical Physics* by Landau and Lifshitz, particularly in the volume *Electrodynamics of Continuous Media*, chapter *Rayleigh scattering in gases and liquids*, see pages 73 and 74. Placzek's contribution is only mentioned as a footnote.

Placzek's Jerusalem anabasis, 1935[13]

June 1934: While in Berlin, Placzek was given a preliminary offer to work as a Professor of theoretical physics at the newly established Hebrew University in Jerusalem. Very likely, Ornstein was instrumental behind this nomination – recall that Placzek worked in his laboratory during his postdoc days in Utrecht. A goal of the Hebrew University was to establish a top department of theoretical physics. Theoretical

---

[12]For further details see [20].

[13]G. Placzek was at the Hebrew University from January till September 1935, see page 227.

physics seemed to the University as "particularly convenient from the point of view of a Jewish nature," and also because – due to opinions of the University officials - "a theoretician needs just a pen and paper." A theoretical physicist of a high international stature was to be considered for the new Professor's chair. The candidates to be approached appeared on the first list as follows: Placzek, Heitler, London, Nordheim. In the course of several years to come, some quite complicated and generally unsuccessful negotiations with several candidates were carried out. Those approached were Georg Placzek, Felix Bloch and Eugene Wigner, but very finally, the chair was obtained by Giulio Racah [22].

The first week in October 1934: as an intermezzo to his Jerusalem adventure, Placzek attended the International Conference on Physics in London, which attracted some five or six hundred physicists from almost all European countries and the USA. A big group photo from this conference shows almost all prominent physicists of that time. For Placzek, it was an ideal meeting to shape his plans.

Discussions about Placzek's nomination were uneasy and lasted in Jerusalem until November 1934, when a formal offer was given to him. Placzek's distinctive preparations for his trip to Jerusalem, which he undertook in Copenhagen, contributed to a portion of those funny stories known as "Placzek's Geschichten." In his recollections, Otto Frisch describes a vivid image of a desperately packing Placzek, who alternately finds and loses his delicately classified items. The story culminates with a crazy drive by a taxi to the railway station. During the wild drive Placzek swears that he will never be late anymore. Fortunately, a conductor of the international train took notice of one passenger (Placzek) missing and delayed departure of the train by two minutes.

January 1935: Placzek eventually reached Jerusalem, where he negotiated with the leadership of the University (Magnes) about conditions of his employment. Both sides faced some unwelcome surprises. Some academic officials were not willing to cap the Professorship so quickly, supposing that having a 29-year old physicist might provoke some older Professors. On the other hand, Placzek was presuming a rather high salary and some funds for traveling around Europe due to some premature promises. He also required a top assistant, and Edward Teller was his choice. Moreover, he demanded an independent office for the theoretical physics department. He also wanted to establish an experimental laboratory for nuclear physics, with an annual fund of 27,000 pounds. Placzek suggested raising these funds with the help of a campaign by Jewish physicists in America – he pointed to Compton, Franck, and Einstein. The academic officials were horrified and their conclusion was: "Placzek's requirements are absolutely unacceptable." An exchange of letters followed, where both sides were formulating their requirements and offers.

On June 11, 1935, another round of negotiation between Placzek and the University Chancellor Magnes took place. Both sides stated ultimately that their requirements were incompatible. Placzek said that he "wanted to build-up, but only static conditions are offered to him." He suggested Bloch, Wigner, or Peierls as possible candidates. He planned to spend some time in England, then Copenhagen, and perhaps America. He presumed that getting an adequate work place should not be difficult. And then Placzek, the grandson of a Moravian Rabbi, sends a telegram to Frisch [22; 5]:

Mit den Juden bin ich ein für allemal fertig.[14]

This "Jerusalem episode" of Placzek's life illustrates the incredibly high reputation which this young physicist (at that time less than 30) enjoyed in the physical community. It also shows that Placzek knew his value well and did not court popularity with academic officials. Moreover, the cosmopolitan Placzek shows here something of his attitude towards Orthodox Judaism and Zionism – indifferent if not negative. Naturally, it was a rather strong handicap in the eyes of the officials of the newly established Hebrew University. Placzek was also opposed to the requirement of the University to lecture on physics in Hebrew – though there is no doubt that he would manage it well – taking into account his renowned language skills.

The confrontation is framed by the fact that the Hebrew University finally payed Placzek only 300 pounds, rather than the 400 pounds initially promised. Statement of the reasons explains why he spent only eight months in Palestine, instead of the whole year. Placzek hardly obtained any scientific benefit from this anabasis, as instead of physics and lecturing he faced nonproductive negotiations with academic bureaucracy.

After his stay in Jerusalem, Placzek spent some time in the USA, as he previously planned. On April 27-29, 1936, he attended the Second Washington Conference on Theoretical Physics. The conference was

---

[14]Done with Jews once and for all.

co-organized by George Gamow and Edward Teller, under the joint auspices of the Carnegie Institution of Washington and the George Washington University. The topics of the conference were as follows:

1. Chemical bond;
2. Reaction velocities;
3. Magnetism;
4. Van der Waal's forces;
5. Molecular vibrations;
6. Isotopes.

This conference is mentioned as a milestone in the history of science because of Hans Bethe's "carbon cycle," a chain of nuclear reactions which was proposed (together with proton-proton reaction) as a likely source of energy of the sun and stars.

Placczek and Landau in Kharkov, 1936-37 [15]

Late 1936: Placzek packed his luggage in Copenhagen once again, this time to make his second trip to visit Landau in Kharkov – together with Victor Weisskopf. Though this trip generated some funny "Placzek Geschichten" again, the outcome of this journey was eventually dramatic and of great historical importance.

As Weisskopf recollects [1], neither he nor Placzek were holding any great illusions about the Stalin Soviet regime at that time, but reality appeared even much worse: a low-spirited atmosphere everywhere, their friends scared and unwilling to talk about anything sensitive. One evening in Kharkov, Placzek was asked under what conditions he would accept employment at the Ukrainian Physical Technical Institute (UPTI). Placzek formulated five conditions, quite similar to those from his Jerusalem anabasis: a sufficient salary – partially in US dollars, a trip to the West every year, a free choice of collaborators, a laboratory, and also that "khozyain" should disappear first. By the Hebrew word "chasain,"[16] which means something like "chaser" or "warder," he implied Stalin. A communist [17] denounced them and Placzek had to flee from the Soviet Union in early January 1937. He was still very lucky.

---

[15] Placzek's second visit to Kharkov lasted from late November 1936 till early January 1937; for more details see page 77.
[16] For a different interpretation of *khozyain* see page 75.
[17] Barbara Ruhemann, see page 291.

On January 24, 1937, Alex Weissberg, who arranged Placzek's trip to UPTI in Kharkov, was summoned to NKVD[18] in connection with the Placzek "affair." On March 1, 1937, he was arrested in Kharkov and accused as a "German spy." He survived several years in Stalin prisons and later published [23] an extraordinary testimony[19] *The Accused* – which begins from the Placzek "affair." Additionally, a Hungarian physicist in Landau's group, László Tisza, avoided imprisonment only by a dangerous escape from the Soviet Union. On February 8, 1937, Landau requested[20] to be accepted as a scientist in the newly established Kapitza's Institute of Physical Problems in Moscow – rather his politically forced escape from Kharkov. And then the "Great Purge" of Stalin and his "cruel dwarf" Yezhov hit the Soviet physicists with all its violence.

On August 6, 1937, an ingenious theoretical physicist Matvei Petrovich Bronstein was arrested in Kiev, just after his return from a summer holiday with Landau in Crimea; he was executed on February 18, 1938 [24]. Very likely, Bronstein significantly influenced the first issues of some early volumes of the *Course of Theoretical Physics* by Landau and Lifshitz, (in particular, the first issue of *Statistical Physics*, issued in 1938). The idea of a comprehensive course of theoretical physics came to Lev Landau and Matvei Bronstein in the late 1920s when both were at Leningrad Physical Technical Institute. In 1932 Landau left Leningrad for Kharkov. Since then they tried to work out this idea sporadically and independently in the 1930s. A copy of the manuscript of *Statistical Physics* (circa 1933) with M. Bronstein and L. Landau on the title page was preserved by one of the graduate students of that time. On the other hand, the title page of typewritten *Lectures on Statistical Physics* dated 1935 and authored by Landau and Lifshitz was found at UPTI.

It is beyond the scope of this article to analyze these sensitive questions of authorship here, including Placzek's contribution to some works of Landau. We also do not know if Placzek and Bronstein ever met. We know, however, that Placzek did visit Leningrad and that Bronstein's

---

[18]Abbreviation for the Soviet Political Police in 1922-1923. This name changed many times without changing its essence – the organ of repression and implementation of the regime of brutal dictatorship. In 1917-1922 it was Cheka, in 1923-1934 OGPU, in 1934-1946 NKVD (with the exception of 1943 when the Soviet Political Police was renamed as NKGB), in 1946-1953 MGB, and, finally, in 1954-1991 KGB.

[19]For more details see [20].

[20]In fact, Landau left Kharkov around January 15. He mailed his letter of resignation to UPTI from Moscow, a few days after his arrival.

very last paper (issued in 1937) is devoted to the scattering of neutrons in magnetic field – a suspiciously Placzekian topic.

From Placzek's second trip to Landau in Kharkov, no clearly authored joint work exists. However, in *Collected Papers of L. D. Landau* [21] one can discover a seven page article of 1937, *On the Statistical Theory of Nuclei*, in which he refers to an article by Bethe and Placzek [25] from the same year – adding a footnote:

> The author would like to express his thanks to G. Placzek for communicating the results of this work prior to publication.

And in the very last sentence of the paper Landau states:

> Finally, I should like to thank G. Placzek for the very interesting discussions of this problem.

It is a demonstration (by far not singular) of the way in which not only Landau, but some of Placzek's other colleagues and collaborators (Frisch, Bohr, etc.) treated his coauthorship. Placzek's influence on stimulating – and maybe also writing – Landau's paper is more than likely (Landau himself practically never wrote his papers, leaving the writing process to his coauthors. It is said that his most extensive manuscript was his "confession," written in prison by his own hand) [24].

This second trip to Kharkov was obviously critical to Placzek's political and philosophical opinions in later years of his life. After these first-hand experiences, he could hardly cleave to the naive ultra-leftist or quasi-communist (or whatever it was) ideology of his youth. Placzek (and also Tisza and Weisskopf) would soon after play an important role in informing later "fathers" of nuclear bombs (Oppenheimer, Teller, Bethe, Wigner, Szilard and others) about the true nature of the Soviet regime. It soon became obvious to this group of physicists that neither Hitler nor Stalin could win the race of the nuclear weapons.

The historical impact of Placzek's role is very underrated and sometimes not recognized at all. Only lately have investigations shown his importance in the proper light. Placzek was particularly convincing in this role of a "lamplighter" – perhaps best of all physicists of his time. It was not only a consequence of his extreme and extensive globetrotting, but also his close and informal personal contacts, and his integrity and skills of persuasion. He was generally admired for his deep insight, his knowledge of languages, and his eloquence. These features complemented his erudition in theoretical, experimental, and mathematical physics. Combined with natural skepticism and a critical attitude toward the work of his colleagues and himself, this rare combination of personal qualities allowed him to grasp the essence of problems before the others. Georg Placzek lives in the memories of his contemporaries as a unique and impressive personality. At the same time, his scientific life shows qualities often neglected by formal scientometry, a fact exemplified by the fate of Georg Placzek.

The most frequently cited unpublished paper

Before leaving for Paris, Placzek spent the first half of 1938 with Niels Bohr in Copenhagen,[21] where he studied with Niels Bohr and Rudolf Peierls the compound nucleus and generic nuclear reactions in the continuum spectrum. As a result, they formulated a theorem relating the total scattering cross section to the imaginary part of the forward scattering amplitude. The corresponding formula, known as the optical theorem, had been known for a long time in optics. However, its emergence in the problem of quantum scattering came as an unexpected surprise. Currently it is given in any textbook in the chapters devoted to quantum scattering. Most often it is referred to as the optical theorem. Only a few most knowledgeable authors mention Bohr, Peierls and Placzek in this connection.

Bohr, Peierls and Placzek drafted and redrafted a detailed paper many times. Apparently, some of them were perfectionists (guess who?). In June 1939 Bohr wrote to Peierls that "Placzek suggested, it would be nice if a short account of the result could appear in *Nature* in the near future, and I shall bring with me a draft Placzek and I have written."

Perhaps, he (Placzek) anticipated the imminent war?

This draft after a minimal polishing was published in the July 29, 1939, issue of *Nature*[22] in the Letter to Editor section, with the

---

[21]Placzek left for Paris in May or June 1938. In his letter of August 27, 1938, Niels Bohr writes to Rudolf Peierls: "I hope you have received my telegram and that it will not be inconvenient for you to postpone your visit here a few days. The matter is that Rosenfeld and I, according to the urgent request of the Organization Committee, are in these days working on the completion of my Report to the Warsaw meeting and that it would be most important to have that off my hands before we start energetically *on our paper on nuclear problems.*" -Sabine Lee, Vol. 1, p. 616.

[22]See N. Bohr, R. Peierls and G. Placzek, "Nuclear Reactions in the Continuous Energy Region," *Nature* 144, 200-201 (1939). One of the surviving drafts of the "long paper" which became "the most frequently cited unpublished paper" can be found in *Niels Bohr's Collected Works*, Ed.

following footnote:

> The details of this and of the other arguments of this note
> will be published in the Proceedings of the Copenhagen
> Academy.

In this letter in *Nature*, Bohr, Peierls and Placzek focused on the photo-electric effect and photo-disintegration. Some of the more general considerations were briefly mentioned.

The detailed paper never appeared. The authors did not manage to finalize a "perfect" text before the outbreak of the war. During the war they were preoccupied with bomb-related work, and after the war Bohr, Peierls and Placzek discussed the issue and decided that, since their result was already common knowledge, there was no need for publishing the promised detailed version.[23]

In the 1969 interview [24] Rudolf Peierls recollects:

> We started writing a bigger joint paper, which has frequently been quoted but never published. The point was that the three of us – Bohr, Placzek and I – had rather different attitudes. I was trying to bring in a lot of the formalism which I had just developed with Kapur, because it allowed one to make concrete quantitative statements on certain things. Bohr preferred rather more general philosophical arguments. And Placzek was sort of halfway. He was interested in our formal results but also interested in trying to pin the arguments down physically. So, as did any paper which involved Niels Bohr, this went through many drafting stages and partial drafts and correspondence back and forth [...]
>
> Finally the war intervened and cut off any communication, at least with me and Bohr. [...] We had talked with many people, of course, about some of the contents, including what was then called the optical theorem, which was a very general result about scattering problems, and which is part of what is now called the unitarity relation – it is sometimes still referred to as the optical theorem

---

R. Peierls, Volume 9 (Amsterdam: North Holland, 1986) p. 49, note 86 and p. 50, notes 87-89.

[23]See their correspondence on pages 168, 169, 180, 190 (February 6, 1948), 191, and 192.

[24]Charles Weiner, Interview with Peierls for AIP Oral History Project, Session II, August 12, 1969, https://www.aip.org/history-programs/niels-bohr-library/oral-histories/4816-2.

which was very satisfyingly general [...] We were quite pleased with that result, which I think arose mainly in discussions between Placzek and myself. So, as a result, later when people wrote up the literature, they quite often referred to this paper by the three of us as an existing paper or as a paper in the course of publication. But, in fact, when after the war there was a possibility of getting together again, the subject had developed so much further embodying all the results of our paper, that there was no point in trying to write it again.

Discovery of uranium fission, the threat of atomic bomb

Now we enter the most strained period in the life of Georg Placzek. As a physicist, he will experience his "finest hours" and become a catalyst of several fundamental discoveries of nuclear physics. Some of his colleagues (e.g. Frisch and Bohr) will mostly impound his contributions, though. Personally, he will be facing the oncoming tragedy of his family in Brno.

June 1938: George Placzek and Victor Weisskopf were traveling together to the West Coast of the USA, heading for Stanford. They spent a week full of late night discussions with Robert Oppenheimer, at his ranch Perro Caliente in the Pecos Valley near Los Alamos (New Mexico). A silent witness to this meeting is a letter dated July 17, 1938, which Placzek sends from the ranch to Rudolf Peierls (see page 157). In this letter, which survived in Peierls' archive, Placzek is inquiring if anything reliable is known about Landau's fate. This meeting with Oppenheimer had some history-making consequences. Personal testimonies of Placzek and Weisskopf about the "Great Purge" in the Soviet Union and true Soviet reality were so fresh, convincing and trustworthy that from that moment on Oppenheimer kept no illusions about the Soviet regime – impacting his attitudes and actions during the Manhattan project and afterwards [1; 26; 27; 28].

In a letter [28] associated with his hearings in 1954, Oppenheimer described the significance of their meeting as follows:

In 1938 I met three physicists who actually lived in Russia in the thirties. All were eminent scientists, Placzek, Weisskopf, and Schein; and the first two have become close friends. What they reported seemed to me so solid, so unfanatical, so true, that it made great impression; and it presented Russia, even when seen from their limited experience, as a land of purge and terror, of ludicrously bad management and of a long-suffering people.

Figure 1.13   Left to right: L. Nordheim, H. Bethe, W. Heisenberg, H. James, G. Placzek, K. Lark-Horovitz, and S. Korff in front of the Physics Department building, Purdue University. Summer School of 1939. Courtesy of AIP Emilio Segrè Visual Archives.

We can hypothesize what would happen if Oppenheimer had kept his pro-communist pro-Soviet sympathies during his scientific leadership of the Manhattan project. At least one hidden communist in his close circle, namely physicist Klaus Fuchs, was already sharing some top-secret nuclear information with the Soviet Union.

January 7, 1939: Prior to his departure for America, Niels Bohr was informed by Otto Frisch that together with Lise Meitner (his aunt), they were going to explain some recent experiments of Hahn and Strassmann's in Berlin as a new type of nuclear reaction – a fission of uranium. Bohr reacted with excitement: "Oh, what fools we all were! Yes, this is how it has to be!"

Immediately Frisch called Meitner to Stockholm and they agreed to prepare their article for *Nature* as soon as possible. However, their argumentation was not quite convincing. Frisch discussed that with Placzek, during their time in Copenhagen. Placzek was skeptical of Frisch's interpretation, but he had fundamental advice for him: he showed him a relatively simple experiment which might prove or disprove the uranium fission in a straightforward way. As Frisch later admitted, he was unaware of Placzek's idea before their conversation [19; 10; 29; 7]. Frisch delayed publication of his article with Meitner. The experiment due to Placzek's idea was not particularly difficult and Frisch performed it quickly on January 13, 1939. Immediately he wrote an article under his name and submitted it to *Nature*. It is strange that Frisch did not list here Placzek as a co-author, and even omitted any note about his contribution. Though much later Frisch did point out Placzek's contribution to "his" important experiment explicitly, this old slip was never corrected properly. Consequently, only a few historiographers of physics are aware of Placzek's true role in the direct experimental proof of nuclear fission.

January 16, 1939: Bohr arrives in New York on board the SS Drottningholm. While sailing, he discussed with Leon Rosenfeld the "new nuclear reaction" due to Frisch's and Meitner's interpretation. Then Wheeler learns about the U-fission from Rosenfeld on their way to Princeton by train. The news spread among a number of American physicists, but it did not immediately reach Fermi (though he was awaiting Bohr in the New York harbor). The articles by Frisch and Meitner, and Frisch alone appeared in *Nature*. In forthcoming days, some groups in the USA repeated the relatively simple experiment of Frisch, confirming the U-fission. Bohr, with all his authority, only cared about the priority of Frisch and his Institute in this discovery. He even clashed with Fermi about it, on the basis that Fermi omitted to cite Frisch during his lecture by the end of January [29].

February 3, 1939: Placzek meets Bohr in Princeton, their first meeting in the USA. A consequence of their discussion, which took place during their breakfast, was the discovery of the role which the isotope Uranium-235 plays in nuclear fission. At the beginning, Bohr relished that after the discovery of nuclear fission, they had no more problems with transuranium elements. But skeptical Placzek insisted that a principal problem remained – there was no explanation for a dependence between the neutron energy and an effective cross-section of the fission. After a few minutes of hectic considerations at the blackboard, Bohr returned with an explanation that he presented to Placzek, Rosenfeld, and Wheeler. "The fission that was observed at low neutron energies, said Bohr, should be attributed to a rare isotope Uranium-235." According to Pais, it was the last great contribution of Bohr to nuclear physics [29]. It is hard to estimate how much this discovery was stimulated by Bohr's discussion with Placzek and the key question Placzek posed [10; 29; 7]. In the article for *Nature* that Bohr wrote immediately on February 7, 1939 in Princeton, there was no mention of Placzek – neither concerning his contribution to the experimental proof of nuclear fission, nor about his contribution to elucidating the role of Uranium-235. "Placzek was helpful in formulating theories of fission" – John Wheeler explains in his first-hand testimony much later [10]. No wonder that even some historiographers of nuclear physics are unaware of Placzek's role in these important discoveries. For example, in the otherwise very detailed and excellent biography of Leo Szilard, the above event is described as well, but without mentioning Placzek and his contribution [30].

June-July 1939: Georg Placzek together with some other physicists tried to persuade Werner Heisenberg not to return to Germany from his two-month stay in the USA. They argued that he inevitably will be forced to develop the atomic weapons for Hitler. But Heisenberg refused to listen. His argument was that – being a true German patriot – he should help his country to get rid of Hitler [12]. On a historical photograph from those days, Placzek, Heisenberg, Bethe and several other physicists are featured at Purdue University (Fig. 1.13, page 34). Tension in the atmosphere is tangible from the faces of Placzek and Heisenberg on the photograph. Unfortunately, Heisenberg was not the only one whom Georg did not succeed to persuade about the utmost danger and hard consequences of Hitler in power.

Tragedy of Placzek's family

Georg Placzek was well aware of the danger of Nazism. He was in daily contact with highly experienced people, such as Niels Bohr, who could foresee political developments. As early as in 1934, Placzek argued with his friends in Copenhagen that Hitler would provoke a war within five years – and even made a bitter bet with Weisskopf: 50 Wiener Schnitzel were at stake. A natural question arises: did Georg make an attempt to forewarn his family in Czechoslovakia personally? We have oral testimonies that he did so. Georg traveled to Czechoslovakia (Brno, Alexovice) to persuade his family to emigrate. Also in one of Placzek's letters to Peierls, written in Copenhagen on May 16, 1938, (see page 156), he writes of his plan to make a short visit to Prague on Friday 20th (and from there to Paris). It was shortly before his departure for the USA in summer 1938 (confirmed by his meeting with Oppenheimer). Obviously, the very last chance for Placzek to visit his family in Czechoslovakia was by the end of 1938, or possibly around mid-January 1939. We know that he was with Frisch in Copenhagen around the New Year in 1939, and then – since February 1939[25] – in the USA during the war years.

Unfortunately, even Georg's renowned eloquence was not sufficient to persuade his family to undertake a quick emigration. Together with some duties towards their factory, Georg's parents – like so many other Jews – simply could not realize how rapidly and dramatically the situation would worsen in Czechoslovakia. Only his brother Fritz escorted his pregnant wife Edith to the safety of England.

Unfortunately, Fritz returned from England to Alexovice – maybe just a few days before the Nazis invaded Czechoslovakia on March 15, 1939. Perhaps, he also wanted a chance to persuade his parents to escape from Hitler. It was a desperate and unfortunate attempt – for he could no longer escape himself...

March 23, 1939: A heavy quarrel was witnessed between Fritz Placzek and some Nazis who emerged from his factory staff (persons named Schmirgel and Hasha were blamed by some contemporary witnesses). Testimony indicates that they trapped Fritz in his office, to prevent him from escaping Alexovice. Fritz Placzek, badly shaken, committed suicide by a shot to his mouth. His son Tony, today the closest relative to Georg Placzek, was born mere days before, on March 19, 1939, in England. We presume that Georg learned about this first family tragedy rather quickly, in the spring of 1939, in those hectic days when a real danger of nuclear chain reactions and atomic bombs was immediately obvious to Szilard, Teller, Peierls, Bohr, Frisch and some other physicists.

---

[25] See page 92.

Figure 1.14   Georg's parents Alfred and Marianne Placzek in the 1920s. Courtesy of J. Kocourková and F. Anthony Placzek.

October 30, 1939: Georg's mother Marianne mails a letter of 10 pages from the Flora hotel in Prague (Fochova 121, today Vinohradská street), addressed to her friend Paula (Scherbak) in England. We learn many things about the fate, thought processes and characters of Georg's parents and his sister Edith at that time. Marianne writes about her concerns for Georg, there had been no message from him since August 1939, not even on his father's birthday (September 11). Thus she wrote to Prof. Bohr, who answered by return – Georg is well – we have been cooperating just recently in the USA. Marianne was very proud about this letter from famous Prof. Bohr, with his own autograph. She wrote a dispatch to Georg to the USA – and Georg responded immediately: "I am all right, a letter is on the way."

This confirms that some contact between Georg in the USA and his family in Czechoslovakia was still possible and active at least by autumn 1939. Marianne also writes that Alfred Placzek is trying to arrange their emigration. Edith is not living with them in the Flora hotel, she is teaching German in Prague. Mother Marianne is planning to learn typewriting, to work as a secretary. She also had returned to Brno to settle Fritz's funeral matters... A remarkable letter, written in

Figure 1.15   Georg's brother Fritz Placzek as a hunter, 1930s. Photo taken in front of Placzek's villa in Alexovice. Courtesy of J. Kocourková and F. Anthony Placzek.

a very calm and stoic style, without any complaint about their cruel fate, it is a witness of Marianne's optimistic and resilient character.

Autumn 1939: two children of Georg's cousin in Alexovice – Erika and Daisy Türkl   are rescued thanks to altruism, bravery, and fore-vision of Nicolas Winton [26] – a young British broker working in Prague

---

[26]Sir Nicholas George Winton (1909-2015) organized the rescue of 669 children from Czechoslo-

who succeeded in arranging 8 (9) trains to carry children from Prague to England. His *Kindertransport* rescued 669 mostly Jewish children from their doomed fate in the Nazi death camps. Today, these children are part of the story of the world-famous "Winton's list" – a story discovered by BBC just only in 1988.

Figure 1.16 Georg's sister Edith on the eve of WWII. Courtesy of J. Kocourková and F. A. Placzek.

---

vakia on the eve of the Second World War: he found homes for the children and arranged for their safe passage to Britain.

One of the last extant memories on Georg's parents and his sister is a postcard dated June 17, 1940, which Marianne sent from Prague to the family of J. Kocourková in Alexovice. It is written in Czech, she thanks them for some foregoing well-wishes.

Next came the transports to concentration camps. Georg's sister Edith appeared in the second transport from Prague to Terezín,[27] with a departure date of November 30, 1941. She was assigned №594 in transport "H." On January 15, 1942, she was put to the transport "P" under №999, northeast direction Riga, where all traces of her disappear. Thus she never met her parents again. Alfred and Marianne Placzek appeared in transport from Prague to Terezín departing on May 15, 1942. It is the transport "Au-1," where they were assigned №582 and №583. In Terezín they were separated. Alfred Placzek died on September 23, 1942. Some testimonies indicate that he starved himself to death.[28] After the war, one Ivančice woman that was rescued from Terezín told that she saw a starving Marianne there, she was asking about some peelings. Marianne was then transported from Terezín to Auschwitz (Oświęcim) on October 12, 1944, where all her traces disappear a few months before the concentration camp was liberated (on January 27, 1945). From Placzek's house in Alexovice, a car full of valuables was taken somewhere to Stuttgart by the Nazi chief of the factory, Reinhard.[29]

Mr. Kocourek, former private driver of the Placzeks, was forced to drive the car (as Ms. Kocourková told us, he pretended inflammation of his eyes to escape the duty, but it did not help). Just after the War, the Soviet Cossacks occupied the Placzek's house in Alexovice with their horses. What little remained in the house was "communized" by some local citizens [16].

---

[27] Theresienstadt concentration camp.

[28] A rare testimony about Alfred and Marianne Placzek's fate in Terezín was preserved thanks to a Czech-American architect Norbert Troller (1896-1984) (recorded by Dr. Ruth Davis, available from the Leo Baeck Institute Archives, Paul and Susan Stern Collection, www.lbi.org). Marianne Placzek was officially pronounced dead on April 10, 1945, though she probably died at the end of 1944.

[29] Oskar Reinhard from Mannheim was one of the three "aryanizators" of the Placzeks' factory *Skene & Co.*, together with Herbert Adolff and Karel Kaess. The aryanization took place on December 11, 1941, when Reinhard became the factory director, succeeding the regent (*Treuhänder*) Ing. Adolf Boček.

Manhattan project: Cornell, Montreal, Los Alamos

In the beginning of the war, Georg Placzek lectured for about three years at Cornell University in Ithaca, New York (probably starting in February 1939). He was there with his good friend Hans Bethe, with whom he published an article *On the Interpretation of Neutron Measurements in Cosmic Radiation*. The article was submitted on November 23, 1939, and appeared in *Physical Review* in April 1940. Victor Weisskopf often visited them and they discussed the atomic bomb threat. Emilio Segrè recalls from that time how his "dear friend George Placzek" consoled himself for having passed from the state of a wealthy gentleman to his present penury: "See! I am at Cornell University. I have an excellent "salary": $1,000 a month, but expenses kill me!" He then listed the "expenses": $150 a month for having escaped the Nazis; $150 for living in a good climate; $100 for the use of the library and for having access to seminars and to worthy colleagues, and so forth. "I am left with only $120 per month, but it could be worse," Placzek concluded [35].

And then came the years when top secret reports authored by Georg (regarding nuclear war projects) became the major written documentation of his scientific work. These reports were only made public several years ago. Georg Placzek was among the first to arrive in Montreal in the end of 1942 to head the Montreal Laboratory Theoretical Division.[30]

January 1943: Georg Placzek was in Montreal by that time, recruiting with all his energy, physicists for a theoretical group under his leadership (see page 173). This group would be developing an Atomic Energy Project in Canada. All testimonies are unequivocal that Placzek proved himself a very impressive, capable and inspiring leader, on both scientific and interpersonal levels. Though there were several capable theorists in the Montreal group, only Placzek had the prestige that allowed him to interact on equal basis with the key physicists of the Manhattan project [11; 12].

October 21, 1943: Placzek mailed a brief letter from Montreal, addressed to Niels Bohr in London. The letter corresponded with a dangerous, but lucky escape, of Bohr and his family from occupied

---

[30]Bengt Carlson recollects: "[I was] in charge of a computing set-up under George Placzek, head of the Theoretical Physics Division. I had been interviewed by Placzek just before Christmas, 1942, in a hotel near the Grand Central Station in New York City." In the same document we can read: "It was Weisskopf who advised Ernest David Courant to join the Montreal project under his friend George Placzek. Courant spent 3 years, 1943-1946, working with Placzek..."

Copenhagen (on September 29) to London (October 6). Placzek was pleased by that and continues: "During all these years I have often been thinking of our last discussions at Princeton in the beginning of 1939, and how right you were then, not only about 235, but also in your political optimism about the future."

December 3, 1943: An international group of (mostly British) scientists joins the Manhattan project officially by the act of the "Toronto protocol," signed by Churchill and Roosevelt. Among the prominent members of this "British Mission" are Georg (George) Placzek, Niels and Aage Bohr, Otto Frisch, Rudolf Peierls, and – alas – Klaus Fuchs (a Soviet spy disclosed after the war). A "Bohemian Placzek, who made general Leslie Groves unhappy," worked in his Montreal laboratory instead of Los Alamos (later this laboratory moved to Chalk River near Montreal). During all of WWII, Placzek was classified as a British citizen. He obtained US citizenship on June 11, 1945 [31; 32; 33; 6].

Placzek's numerous, extensive, and significant works of this war period belong mainly to the area of mathematical physics. They show the merit in both the atomic bomb project and atomic reactor development in that period. His reports were available only for a limited circle of his collaborators and their wider dissemination during the war was impossible.

There exist referee's reports (by Wigner and Fermi) of one of Placzek's works of that time, namely, *The Theory of the Slowing Down of Neutrons*. They rate the work very important, but they recommend to keep it secret and avoid its publication. During that time, Placzek also developed a very elegant and efficient numerical method for solving integro-diferrential transport equations for the neutron diffusion. This method obtained admiration in Fermi's group in Chicago, because only cumbersome "fist formulas" (as Placzek called them) had been used for the task before [33]. Among other results that were published after the war, it is worth naming numerical evaluations of the so-called "Placzek's function," which is important in diffusion and probability theory. A list of Placzek's research reports is now available in a special issue of *Progress in Nuclear Energy* [11].

M.M.R. Williams writes. "At the time he was working in Canada, Placzek was classed as a British citizen. According to Professor P.R. Wallace, Placzek was an exemplary leader of the theory group in Montreal and had the necessary stature to be on equal terms with all the leading scientists of the Manhattan Project. Aside from his brilliance as a scientist who made contributions in many fields of nuclear physics, he never criticized and was always enthusiastic about the work done by the members of his group and contrived to make everyone feel that they were a vital member of the team."

Figure 1.17   Georg Placzek with his wife Els, circa 1943. Courtesy of M. Fuhrmann.

***

To comply with top secret regulations, Niels Bohr used the pseudonym "Nicholas Baker" during the war. A funny story is credited to Robert Serber:

> Late in December 1943, in a Washington hotel, Bohr entered an elevator and found himself face to face with a

beautiful young lady whom he had often been meeting with before in Copenhagen. "Oh, Professor Bohr, I'm so happy to see you again!" But Bohr, to hide his identity, replied to her: "You must have mistaken me with somebody, my name is Nicholas Baker!" However, to show some politeness, he added: "But I remember – you are Mrs. von Halban!" But the lady answered: "No, I am Mrs. Placzek." As was soon clarified, this beautiful Jewish Dutchwoman, known as Els, had often visited, with her former husband Hans von Halban, Bohr's Institute in Copenhagen. Later she had divorced von Halban and married Georg Placzek in 1943... (see Appendix, page 335).[31]

Els von Halban (Fanny Ella, née Andriesse[32]) met Georg Placzek in September 1937 during a conference that was held in "Congres du Palais de la Decouverte" in Paris. As Lászlo Tisza recollects (see page 226), there were many social events. Els von Halban was one of the stars and Placzek did not conceal his admiration for her. About that time Georg Placzek was collaborating with Hans von Halban, Jr., an assistant of Frédéric Joliot-Curie, at the Collége de France in Paris. Els gave birth to her daughter, Catherine Maulde von Halban, on August 24, 1939 in Boulogne, while Georg was already in the USA. Hans von Halban later became the head of the Montreal laboratory in Canada and suggested Placzek as a leader of the theoretical group. Judging from some of their letters (available in [11]), their personal relationship remained quite friendly. Both their names are engraved on a memorial plaque at the Montreal University, unveiled on May 17, 1962 [11].

Trinity

Georg Placzek was obviously the only citizen of Czechoslovakia who was directly and significantly involved in the Manhattan project. Early in 1945, as the bomb projects were reaching their critical stage, Placzek was assigned to duties at Los Alamos and in May 1945 moved there

---

[31] Some sources [1; 2; 6] indicate that Georg Placzek was married twice. Even if the first marriage took place, nothing is known about his first wife. Weisskopf mentions that his first marriage was unhappy. In contrast, his second marriage (to Els Placzek) was obviously happy.

[32] Els Placzek (1912-2014) had a daughter from her first marriage, Maulde Griffin (1939-1993), née von Halban (Placzek's stepdaughter). In the 1980s and 1990s she lived in Oxford, UK, and Els often visited her there, see Rudolf Peierls' Diary, to be published, and page 335.

from Montreal (together with Carson Mark, see page 349) [12]. It is also highly probable (judging by indirect evidence) that Placzek witnessed the very first explosion of the atomic bomb on July 16, 1945, in the desert near Alamogordo, New Mexico. We know that all top scientists – his good friends from Los Alamos – witnessed the horrifying experiment personally, and it is very unlikely that Georg Placzek would stay away.

In the days after Japan surrendered, in September 1945, the "British Mission" in Los Alamos entertained their guests at the celebratory party with a skit depicting what happened at Trinity near Alamogordo. In the skit's finale, documented on several photographs, a makeshift tower supports a gadget that was detonated with remarkable sound and light effects. And standing beside this "Trinity site" is Georg Placzek [31; 27]. It is another indirect proof that he was not only in Los Alamos, but also witnessed the first atomic bomb explosion near Alamogordo personally.

In Goodchild's book [27] there is a mysterious photograph from the most critical phase of bomb assembly. It shows silhouettes of several scientists under a tent near the historical "Point Zero" – Trinity Site at Alamogordo. The fine operation – entering a "Schem" to the atomic Golem – is accomplished by a Russian specialist on explosives, George Kistiakowski, a skeletal Robert Oppenheimer is watching around, and also... there is a silhouette inclining above the bomb mechanism which resembles spectacled Georg Placzek. Later we verify that it is not him – though there might be some logic behind it: he was one of Oppenheimer's very good friends, he was a theorist who was not scared by experiments, his combinatorial thinking was fast in a case of need. Few persons around Oppenheimer were more convenient for this historical task. Georg also had every moral and philosophical justification for such a step. Of all the physicists involved in the Manhattan project, his family had been affected most by the Holocaust tragedy – a fact surely known among Georg's friends and colleagues at that time.

It is known that Oppenheimer, with his sense for mysticism, had been thinking also about these symbolic aspects of the Manhattan project [26; 27]. In any case, a whole chain of brains and hands were used to put the "Schem" inside the atomic Golem. Including Georg Placzek – a grandson of Rabbi Baruch – who once in Brno inspired young Georg to science.

This section may be concluded by yet another "Placzek's Geschichten." A family legend has it that through his grandmother,

Caroline Löw-Beer and her relatives in Budapest, Georg was linked to the famous Judah Löw, son of Bezalel ("Rabbi Löw" of Prague), the creator of the Golem. And with certainty, through the same Löw-Beer family, Georg Placzek was also linked to the celebrated Tugendhat villa in Brno, and to a textile factory in a small village Brněnec – the stage of the world famous "Schindler's list." Well, as they justly say, Georg Placzek was behind "almost everything."

Years after the war

After Hiroshima, Nagasaki, and the surrender of Japan, the scientific staff in Los Alamos changed dramatically. On December 1, 1945, Hans Bethe resigned and George Placzek replaced him as head of the Theoretical Division. His work status changed too: he was no longer a member of the British Mission, but, rather, was directly employed by the US Government.

On April 23, 1946, he mailed from "Santa Fe" (actually Los Alamos) a significant letter to Niels Bohr in Copenhagen. In this letter, together with an invitation to a conference in Los Alamos, Placzek drew to Bohr's attention new experiments of Felix Bloch and Edward Purcell – addressing the discovery of Nuclear Magnetic Resonance, for which these two scientists would soon be honored by the Nobel Prize. It demonstrates another example of Placzek's remarkable insight. However, the climate "on the Hill" proved not to be good for his health, and on May 20, 1946, Placzek became ill and left Los Alamos soon afterwards.

Some of his relatives recollect that (in spring or summer 1946(?)) he taught at the University in Montreal, residing in a large old-fashioned apartment on the street Côte des Neiges. In 1946-48 Georg Placzek was affiliated with the General Electric Research Laboratory in Schenectady, New York. "Dr. Suits, the Director of the Laboratory, arranged for a spectacular series of seminars for the senior staff, to be delivered by several prominent veterans of the Manhattan Project, including Hans Bethe, Richard Feynman, Edward Teller, George Placzek, Philip Morrison, and several others. The small group of ten or so staff who attended those seminars ultimately formed the nucleus of the Knolls Atomic Power Laboratory" – recollects Harvey Brooks about spring or summer 1946 in his autobiography [34]. "At General Electric, I found Bethe, Placzek, and Pontecorvo" – recalls Emilio Segrè about his visit there in the summer of 1947 [35]. The Placzeks lived in Schenectady at the address 918 St. David's Lane. Els' US citizenship was announced in the *Schenectady Gazette* on July 7, 1947.

Figure 1.18   The 1947 Copenhagen Conference organized by Niels Bohr. Front row; Niels Bohr, Léon Rosenfeld, Wolfgang Pauli, Ralph de Laer Kronig, George Placzek, B. Ferretti, John Archibald Wheeler, Lise Meitner, and Otto Frisch. The famous Auditorium A of the Niels Bohr Institute. Credit: Niels Bohr Archive, Copenhagen.

In 1947, Placzek attended once again his familiar conference, "Bohrfest" in Copenhagen (Fig. 1.18), probably his first visit to Europe after the war. This conference is documented by several photographs in the Niels Bohr archive. We see Georg Placzek, changed by his family tragedy. And, in contrast to the previous "Bohrfests," he is in the prominent front row for the first time, side-by-side with Bohr, Pauli, Wheeler, and other dignitaries.

In 1948, Placzek was invited to the Institute of Advanced Study (IAS) in Princeton, where his friend Robert Oppenheimer had become the director. Some details of Placzek's arrangement with IAS are described by Beatrice M. Stern in her book on the IAS history:[33]

Perhaps because of the financial stringency at the begin-

---

[33]Beatrice M. Stern, *A history of the Institute for Advanced Study 1930-1950*, Part 2, https://library.ias.edu/files/stern_pt2.pdf

ning of its term, Dr. Oppenheimer managed to achieve a measure of deliberation in the recruitment of younger mathematicians and physicists. Judging by past history, this could only have been true if Professor Veblen enjoyed a degree of confidence that none was competing with him in the spending of available funds. Dr. Oppenheimer confined his first efforts to calling in for periods of a semester or a year several distinguished physicists and many younger men as members. He did not move to gather together a small permanent group until later. During his first two years, Dirac, Bohr, E. Hylleraas, Pauli, Hideki Yukawa, G. E. Uhlenbeck, and some younger men of distinction were in residence at various times. In the latter group were Freeman Dyson, George Placzek and Chen Ning Yang. But the first permanent member of the physics staff was Dr. Pais, the young Dutch physicist who had spent his first postdoctoral years in the Dutch underground, where, as a fixed contact point, he devoted most of his time to working intensively in physics, and to good effect. (653)

At the time there were three temporary members in theoretical physics whose work Dr. Oppenheimer found very worthy. They were Freeman Dyson, professor at Cornell, George Placzek, who had worked at Chalk River and Los Alamos, and Dr. Chen Ning Yang. Dr. Oppenheimer was still the only active professor in theoretical physics in 1950 when he began to feel the need for a small group of permanent men in physics. Dr. Dyson had returned to his native England, and the Director feared he was lost to the Institute. But later, Dr. Oppenheimer was to be rewarded for his patience, for Dyson joined the Institute Faculty. Early in 1950 five-year memberships were voted for Dr. Placzek and Dr. Yang. Dr. Oppenheimer was apparently unwilling to accept co-administrator, and made his view clear by advising the Institute to purchase the house. It was then rented for a time to members, and finally sold to Dr. Placzek because it was considered improper for the Institute to rent professors' housing to members. (671-672)

Usually the Director mentioned the appointments of physicists by informing the School of Mathematics of actions of the Physics Committee. The minutes of that same meeting imply a difference between the Committee and himself: "Dr. Oppenheimer discussed the problem of long-term appointments in physics, reporting the conclusions of the Committee on Physics. On the basis of this report, the School of Mathematics endorses the proposal to appoint Professors Richard P. Feynman and Julian Schwinger as professors of physics, provided they are ready to accept such positions, and subject, of course, to faculty and trustee approval. The School of Mathematics also endorsed the proposal to give Dr. Chen Ning Yang and Dr. George Placzek appointments as members for a term of five years with stipends of $5,500 and $9,000 respectively."

Dr. Yang was a brilliant young man who had worked under Fermi at Chicago where he had taken his Ph.D. He had been a member since 1949. Dr. Placzek, forty-five years old, had worked with Bohr, Bethe and Rabi, all of whom commended him highly. He had been at Chalk River and Los Alamos; he was at the Institute first in 1949[34] on an Atomic Energy Commission fellowship. Dr. Oppenheimer said he had found ample evidence of Dr. Placzek's "good and fastidious work in physics." The Director was obviously responsible for the nominations of Placzek and Yang. (681)

That the trouble arose over a difference within the School of Mathematics over the Director's choice of men was to be shown later, when he asked authorization for permanent memberships for them. Then the School of Mathematics legislated, presumably on Dr. von Neumann's motion, that while Dr. Yang might some day become a professor, Dr. Placzek could not, "unless circumstances now unanticipated supervene." The position of the Director was very difficult at this point. (682).

From [36] we learn that Placzek's affiliation with the IAS (Math)

---

[34] Apparently a misprint, should be 1948-49.

officially lasted from September 1, 1948 to June 30, 1955. It is worth noting that he bequeathed his estate to Princeton University.

$$***$$

Report IAS for the Academic Years 1987-1988 and 1988-1989

FUNDS HELD IN TRUST BY OTHERS

The Institute is the residuary beneficiary of a trust under the Will of George Placzek, Deceased, and upon the death of the life tenant will be entitled to receive the corpus thereof. The approximate market value of the assets under the Will, as reported by the administrator of the Estate, aggregated $1,683,000 as of June 30, 1989, and is not included in the accompanying financial statements.

Figure 1.19   George Placzek in Princeton [tentatively, early 1950s]. Credit: AIP Emilio Segrè Visual Archives, Uhlenbeck Collection.

During the hunt for Klaus Fuchs' American contact around 1950,[35] the FBI investigated Placzek thoroughly. His hatred of communists (he called them "animals and liars") and strong support from Isidor Rabi helped clear him of suspicion.

In the famous Princeton's Institute, full of scientific luminaries, Placzek's path miraculously crossed with that of another former student from Brno and Vienna, namely, Kurt Gödel, the celebrated logician, philosopher and occasional physicist. Gödel was born in Brno in 1906, the same school year as Placzek. Unfortunately, there is no evidence of contact in Brno, Vienna or Princeton. We know from several testimonies that under Oppenheimer's directorship, the physical and mathematical departments were kept rather apart and separated, a situation considered as very unhealthy for the Institute. As an exception, Gödel and Einstein were quite good friends and even cooperated on some aspects of general theory of relativity.

With a stoic mind, strong will, and help from his wife Els, Georg faced his burgeoning health difficulties.[36] His windows covered with dark curtains prevented any daylight from entering when his illness struck and paralyzed him for ever longer periods. In spite of these difficulties, in 1952 he published one of his most cited[37] papers: *The Scattering of Neutrons by Systems of Heavy Nuclei*. In cooperation with K. M. Case and F. de Hoffmann he also wrote *Introduction to the Theory of Neutron Diffusion* (released in 1953), presenting a collection of the lectures which Placzek had been presenting in summer 1949 at Santa Monica and Los Angeles. Together with Léon Van Hove, his final collaborator, he published his last papers, including *Crystal Dynamics and Inelastic Scattering of Neutrons* (1954). The spring and summer of 1953 sees Placzek in Europe again, presenting a series of lectures on neutron diffusion and slowing down at Universities in Rome and Milano, Italy. A photograph from this stay exists in Amaldi's archive at La Sapienza University in Rome. Placzek, dark sunglasses and a cigarette as usual,[38] is depicted here on a trip to a castle in the Frascati

---

[35]Klaus Fuchs was a member of the British Mission at Los Alamos Laboratory, and, simultaneously, a Soviet spy. His courier in New Mexico was Harry Gold. After the war Fuchs returned to the UK where he was arrested in January 1950.

[36]Recently obtained documents of Placzek's legal autopsy (Zürich, 1955, see Appendix, pages 343 and 344) present more details of his medical condition.

[37]See Manuel Cardona, Werner Marx, *Georg(e) Placzek: a bibliometric study of his scientific production and its impact*, https://arxiv.org/pdf/physics/0601113.pdf

[38]Tony Placzek recollects that his uncle was a chain-smoker. Once a doctor told him: "If you do not quit smoking, you will die within half a year." Georg Placzek replied: "If I quit smoking, I'll die tomorrow."

Hills near Rome, with Bruno Touschek, Edoardo and Ginestra Amaldi.

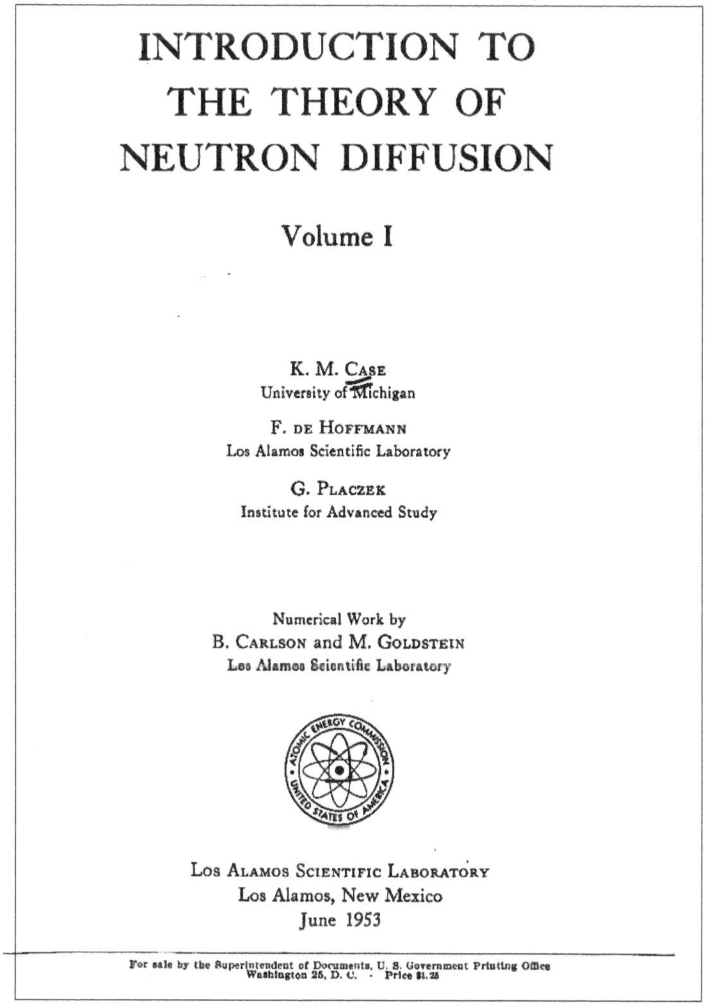

INTRODUCTION TO
THE THEORY OF
NEUTRON DIFFUSION

Volume I

K. M. CASE
University of Michigan

F. DE HOFFMANN
Los Alamos Scientific Laboratory

G. PLACZEK
Institute for Advanced Study

Numerical Work by
B. CARLSON and M. GOLDSTEIN
Los Alamos Scientific Laboratory

LOS ALAMOS SCIENTIFIC LABORATORY
Los Alamos, New Mexico
June 1953

For sale by the Superintendent of Documents, U. S. Government Printing Office
Washington 25, D. C.   ·   Price $1.25

Figure 1.20   The title page of *Introduction to the Theory of Neutron Diffusion*.

In 1955, Placzek obtained Guggenheim's fellowship, planning to spend the academic year 1955-56 at the University in Rome, in the circle of his close friend Edoardo Amaldi. In early spring of 1955 he arrived in Europe. First he went to Heidelberg via Paris, to participate in "Kopfermann-Festspiele" – a colloquium celebrating the 60th anniversary of his good old friend Hans Kopfermann. On April 26th, around 17:15 in "großen Hörsaal des Physikalischen Institutes, Heidelberg, Philosophenweg 12," Georg Placzek ("on leave of absence" from Princeton) presented one of his very last lectures, entitled "Kristalldynamik und Neutronenstreuung." A photograph from this colloquium, printed on April 27, 1955, in *Die Rhein Neckarzeitung*, shows Placzek with William Fowler, David Frisch, Ramsey and Kopfermann. It is the very last known photograph of Georg Placzek [9].

Placzek was seriously ill, however. For some time, he was dwelling with the family of physicist Res Jost in Zürich and when his health condition got worse, he was admitted to a hospital. This is probably where his nephew Tony Placzek, then 16 years old, visited him for the last time. Very likely, this visit preceded Placzek's 50th anniversary. As Tony recalls, his uncle Georg was unable to eat anything or communicate. Just "go go," he told him after several minutes together. On September 25, 1955, the mother of his wife Els died. Els left ill Georg in Switzerland, to attend her funeral in the USA.

Then came Friday October 7, 1955. We can imagine that on this day in Switzerland, Georg may have thought of the anniversary of the death of his grandfather Baruch, whose grave at the Jewish Cemetery behind the "iron curtain" in Brno he could hardly visit. Or did he think about something else? Anyway, Georg left the hospital and moved into a hotel.

In the letter dated October 12, 1955, Wolfgang Pauli wrote to Erwin Panofsky [39] that Georg Placzek was found dead in his hotel room on Sunday, October 9.[40] "On Saturday, added Pauli, Placzek seemed to be well and attended a party." In the same letter Pauli wrote that – seemingly – Placzek was lying dead in his hotel room for several hours before they had found him. Obviously he was then transported to the *Gerichtsmedizin* in Zürich. Pauli concluded this part of the letter by a remark that the first examination detected pneumonia and a weak heart.[41]

---

[39] Erwin Panofsky (1892-1968) was a German art historian of Jewish descent who left Germany for the US after the rise of the Nazi regime.

[40] This statement is not quite accurate. Placzek was found dead on Monday, October 10, and then J. Colombo, MD, established that his death actually occurred 24 hours earlier, on October 9. See pages 342-347.

[41] Courtesy of Professor Ernest Kopp. Professor Karl von Meyenn kindly provided us with the exact reference source: *Wofgang Pauli. Wissenschaftlicher Briefwechsel mit Bohr, Einstein, Heisenberg u.a.*, Ed. K. von Meyenn, (Springer, 2005), Vol. IV/3, p. 367f. The same message was sent by Res Jost to Oppenheimer, as indicated there in a footnote.

The most detailed of Placzek's obituaries by Amaldi [37] indicates that George Placzek died after a long illness on October 9, 1955 (Sunday) in a hospital in Zürich. However, this information is not complete. Recently discovered police medical report sheds more light on the shadows and mysteries of Placzek's death. Let us cite the basic part of this document verbatim:

> On Monday the 10th of October 1955 the 50-year-old physicist Georg Placzek was found dead in his room at the hotel "Im Park." Dr. J. Colombo who pronounced him dead at about 21:00, estimated on the basis of the corpse changes that death should have occurred about 24 hours before. Dr. Placzek had been seen alive for the last time on Sunday evening 9th of October around 7:30 pm.
>
> According to the police investigation Dr. Placzek had been previously treated by a number of physicians, he had already exhibited suicidal thoughts and had to be admitted in closed psychiatric facilities. His bedroom was a mess, which indicated that Dr. Placzek had took his clothes in a hurry. On the table there were several bottles of sleeping pills, including among others Somnifen (two original bottles, empty), a likewise empty pack of Nembutal tablets, various partially consumed partially intact packs of sleeping pills. Among the documents of Dr. Placzek there were found several medical certificates verifying that he had a sympathectomy (surgical removal of vegetative nerves on the both sides of the abdominal aorta) about five years before due to essential hypertension (increase of blood pressure). The operation had been successful and in the last few years Dr. Placzek has had more or less normal blood pressure under common blood pressure drugs. Among the medical prescriptions there were a lot of those for sleeping pills, especially for Somnifen, for Medinal, for Evipan, for Plexonal forte, all dated from June to August 1955.

Today we believe that Georg Placzek had decided about his passing away by himself. It is a scientific fact that bipolar disorder and suicides are highly correlated and genetically predisposed. Half jokingly, a saying about the curse of "Placzek's gene" is used even today among young generations of Georg Placzek's relatives. But among older generations of close relatives, it was no joke at all. Medical textbooks show examples of this strong genetic predisposition on some famous families, e.g., the family of Ernest Hemingway (five suicides in four generations). In Georg Placzek's genealogical ancestry, we find 5 (or probably 6) suicides in three generations, and also multiple appearances of serious psychiatric diseases (clinical depressions). In the 1950s, a new drug called Reserpin was commonly used for treating high blood pressure problems (one of Georg Placzek's problems according to the obituary and the medical autopsy report). However, the Reserpin was later found to have a terrible side effect: it triggered dangerous psycho-depressions.

Particular details of Placzek's final day, October 9, 1955, are now much clearer, though some uncertainties will probably remain forever. The members of his extended family whom the author (A.G.) talked to or corresponded with were unanimous in that Georg committed suicide. His funeral took place in Zürich on October 14, 1955. The funeral oration was given by Léon Van Hove.[42] Placzek's body was cremated on the same day. Many years later,[43] his ashes were buried in a common burial of the Friedhof Sihlfeld in Zürich.

We can just hypothesize to what extent Georg's last years were influenced by undue side effects of the drugs available in his time. And also, it is well known that bipolar disorder and similar diseases are often triggered by periods of stress. During and after the war, Georg could not escape the burden of his family tragedies. It is easy to imagine his twinges of guilt that he was unable to persuade his family in Brno to emigrate. He could hardly escape the dark philosophical and political questions, common to leading physicists of the nuclear era. In 1954, at the peak of the cold war and McCarthyism, Georg Placzek was called to provide testimony at the infamous security hearings of his old friend Robert Oppenheimer. He radically refused any involvement in this matter using the words: "Who do you think I am?" [9].

---

[42]Presented in this Volume on pages 298 and 300; also published in Nuclear Physics, **1**, 623 (1956).
[43]Between late 1989 and 2009. The whereabouts of Placzek's burial place were determined thanks to Marie Fojtíková (Brno) and Martina Weiss (Zürich) in October 2017.

Figure 1.21 George Placzek's late portrait from the obituary by Léon Van Hove, 1955. See page 298 and the subsequent pages.

It is sad that Robert Oppenheimer (whose leadership of IAS was at times controversial both along the professional and personal lines) obviously did not show much gratitude for Placzek's response to his recent security hearings. Such a sad picture is supported by recollections of Freeman Dyson, now Professor Emeritus at the IAS and one of the last witnesses of Placzek's affiliation with the Institute. By the way, Dyson lived in "Placzek's house" in Princeton, and when some strange sounds were heard in this house by Dyson's children, they joked about "Placzek's ghost." "I greatly admired Georg Placzek" – these were the words Dyson used in an oral interview with Karen Kruger in September 2005. In the interview, he also provided a dramatic statement that the mighty Oppenheimer "kicked out" seriously ill Placzek from the IAS and perhaps indirectly contributed to his tragic premature end. Another topic for a historical novel and deeper investigation...

Yes, there are still many shadows, mysteries, and underrated credits around Georg Placzek. It is however quite clear today who Georg Placzek really was: an eminent physicist with a unique style and scholar of world renown, an amusing and inspiring personality with a renaissance knowledge, a humble man with a friendly charm, personal integrity and touching fate, a truly historic and iconic personality of the 20th century.

Figure 1.22   The bronze plaque of Georg Placzek (Brno), created by Nikos Armutidis (the portrait) and Jaroslav Klenovský (graphic design). Aleš Gottvald's archive.

Acknowledgments

A precious memory on his two encounters with Georg Placzek, in Kharkov 1935 and Paris 1938, was conveyed to me by László Tisza (1907-2009), at that time Professor Emeritus at MIT (USA). Though he was already 97, he answered my email within two days! Some obituaries and much appreciated information at the beginning of my Placzek investigations was unselfishly provided to me by Prof. Jan Fischer (Institute of Physics of the Academy of Sciences of the CR), whose truly revivalistic paper of 1985 probably saved important lines of Georg Placzek's story from being totally lost and forgotten. Of living Placzek relatives, invaluable help and trust was provided to me by Michael Fuhrmann (Toronto, Canada), who collected family memories, photographs, and documents. Rare memories, documents and photographs related to their life near the Placzeks were preserved by MU Dr. Marie Nešpůrková with her husband (Alexovice), and by Ing. arch. Jarmila Kocourková (Brno). Encounters with them are unforgettable to me. Many thanks are also due to Mgr. Marta Němečková and Marie Havlíčková (Ivančice Museum) for establishing my contacts with several local historians and contemporaries (Jiří Siroký, Ing. Karel Figer, J. Halůzková). Some photographs and documents from the archive of J. Kocourková were made available to me thanks to Dr. Anita Franková from the Jewish museum in Prague. Placzek's thesis and some obituaries were provided by Prof. Michal Lenc, Dr. Jan Janík, and Dr. Carlos Granja. Ing. arch. Jaroslav Klenovský was helpful in the identification of houses related to Placzek's family in Brno. Fundamental information about Baruch Placzek and his contacts with Johann Gregor Mendel and Charles Darwin was obtained by Dr. Ruth Davis from the Czechoslovak Jewish Communities Archive in New York. Prof. Gerald Wiemers contributed some information about the stays of Georg Placzek in Heisenberg's group at the University of Leipzig. Historically valuable correspondences of Georg Placzek with Rudolph Peierls were provided to me by Dr. Sabine Lee from the University of Birmingham. For some personal memories on four encounters with his uncle, I am very grateful to Mr. Anthony Placzek, nowadays the closest living relative of Georg Placzek. For arranging the interview with Prof. Freeman Dyson in Princeton my thanks go to Karen Kruger (USA). Rare testimony about the last days of Georg Placzek in Zürich was provided by Madam van Hove, and information about his burial place by Dr. Marie Fojtíková and Dr. Martina Weiss (Zürich). I am grateful to all.

My very special thanks and gratitude go to Prof. Mikhail Shifman (University of Minnesota, USA) for his many significant contributions, corrections and suggestions to the text of the present biographical chapter. Thank you indeed, Misha!

Figure 1.23   Tony Placzek and Aleš Gottvald, Brno, October 2017.

Appendix 1: Placzek's Timeline

Born September 26, 1905, Brno

1911-1916(18?)	Elementary school, Brno
September 1918-June 1924	Deutsche Staats Gymnasium, Brno
September 1924-July 1925	University of Vienna (chemistry)
Circa 1925-1926	Three semesters at German University in Prague (physics)
Circa 1926-28	University of Vienna (physics)
July 1928	PhD Thesis defended (Third Institute of Physics, Vienna University)
1928-1930	University of Utrecht (with Kramers and Orstein)
Autumn 1930-summer 1931	Universities of Stuttgart, Leipzig (with Heisenberg), Göttingen, Berlin
Summer 1931-1932	University of Rome (with Fermi, Bethe, Teller)

1932 till mid-May 1938

Bohr's Institute in Copenhagen (with interruptions)

★ 1933, May to December	Ukrainian Physical Technical Institute (UPTI), Kharkov (with Landau)
★ Summer 1934	Berlin
★ First week of October 1934	London, International Conference on Physics
★ 1933, 1934, 1936, 1937	Participated in the Bohrfest Conferences in Copenhagen
★ January 1, 1935-September 30, 1935	Hebrew University of Jerusalem
★ April 27-29, 1936	Second Washington Conference on Theoretical Physics, Washington, DC (with Gamow, Teller, Bethe)
★ Late November 1936 to early or mid-January 1937	UPTI, Kharkov, USSR (with Landau)
★ November or December 1936	Moscow, meeting with Nikolai Bukharin
★ September 1937	Paris (Conference in Congres du Palais de la Decouverte)
May 20, 1938	Prague (one of Placzek's last visits to Czechoslovakia)
1938, late May-June	Paris, Institut de Chimie Nucléaire (with Joliot Curie)
1938, July-September	USA (traveling with Weisskopf from the East to West coast, bound for Stanford, CA)

★ A week around **July 17, 1938**

    Perro Caliente ranch in New Mexico (historically important discussions with Oppenheimer and Weisskopf)

October 1938-December 1938	Paris, Collège de France, Dept. of Nuclear Chemistry (with Joliot-Curie and von Halban)
January 13, 1939	Copenhagen (Placzek-Frisch experiment confirms uranium fission)
1939, Early February	Arrival in the US
February 3, 1939	Princeton, NJ (discussion with Bohr elucidates the role of U-235 isotope in the uranium fission)

February 1939-July 1942

    Cornell University (with Bethe)

★ 1939, June 27-July 1	Purdue University Conference (the last encounter with Heisenberg before WW-II)
July 1942	Teddington, Middlesex, England (Dept. of Scientific and Industrial Research)
1942, August-November	Cavendish Laboratory, University of Cambridge, England

December 1942-April 1945

    Montreal Laboratory (National Research Council of Canada, Head of Theoretical Physics Division)

★ 1942, December 3	Joining British Mission
★ 1942, Late December	Arrival in Montreal
★ 1943	Marriage to Fanny Ella (Els) von Halban (née Andriesse)

May 1945-May 20, 1946

    Los Alamos, NM

★ June 11, 1945	US naturalization
★ July 16, 1945	Alamogordo, NM (Trinity test explosion)
★ December 1945-May 1946	Head of Theory Division (succeeding Bethe)

1946-1948

    Research Laboratory of General Electric in Schenectady, NY

★ Spring or summer 1946	Schenectady seminars delivered by prominent participants of the Manhattan Project (Bethe, Feynman, Teller, Placzek, et al.)
★ 1947, late September	Copenhagen, conference in Niels Bohr's Institute

September 1, 1948-June 30, 1955

    Institute for Advanced Study, Princeton, NJ (with interruptions)

⋆ Spring and summer 1953   Lecturing in Italy at the Universities of Rome and
Milano

⋆ April 26, 1955   Heidelberg (probably Placzek's last lecture, presented at
Kopfermann's Festspiele)

⋆ Summer 1955   Zürich (with Res Jost's family, later hospitalized)

Death   **October 9, 1955**, in Zürich

Appendix 2: Placzek Family Tree

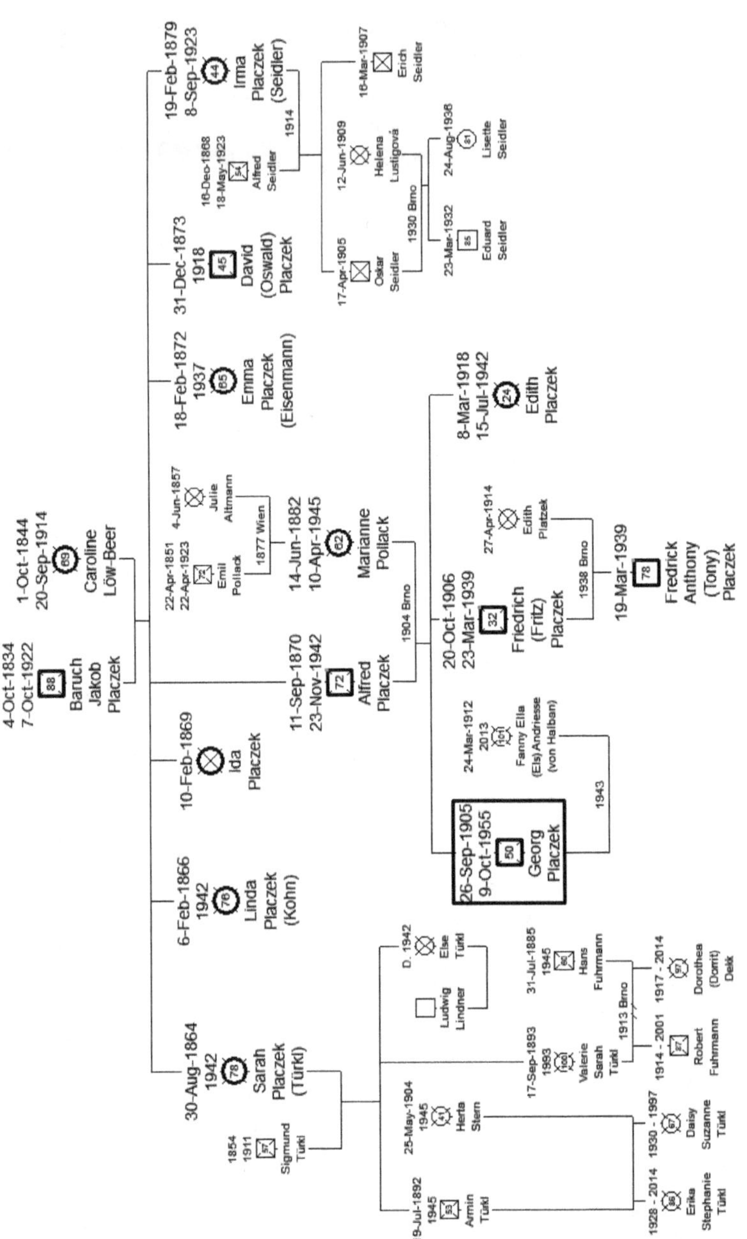

Figure 1.24   Placzek Family Tree.

Human Side of Georg Placzek:
Snapshots from 1930s-40s

Mikhail Shifman

George Placzek, a native of Czechoslovakia who grew up in Vienna, was one of the most extraordinary people I have ever known. We considered him a wise man with whom we were eager to discuss personal, political, and scientific problems. Placzek had an especially clear insight into the problems of the time, whether in physics or politics. But for some reason he was never able to find the right style of life for himself, and his personal life was not a happy one...

As a devout bachelor, Placzek was horribly afraid of the famous, unwritten Bohr Institute Rule: Any physicist working with Bohr was certain to be married after no more than two years. When his second year was nearly over, Placzek began to worry that his happy days as a single man were about to end. Since I was engaged to Ellen after only a year in Copenhagen, I decided to give him the year that was still owed me by tradition. In a solemn ceremony I transferred it to him so that he could be secure for one year longer. Much later when he did wed, the marriage was an unhappy one, so perhaps his instincts in this regard were sound.

Victor Weisskopf [38] [44]

If scientific contributions of Georg Placzek to quantum physics, as well as his role in the Manhattan Project are rather well documented, his personal life is shrouded in the fog of oblivion. All we know are marginal mentions in the memoirs of his friends and colleagues. However, there is one chapter of his life which was narrated by Placzek himself in his letters and the replies of his correspondents. These letters were discovered recently in private archives and are not yet known to the general public.

I remember that when I first came to ITEP in 1970, Placzek's name quite often was pronounced with respect by my PhD adviser Boris Ioffe and other physicists from the ITEP Theory Department (I do not remember the context, though). From Placzek's aura in ITEP I got an impression (which is still vivid) of him as a great physicist. Recently, I looked through Wikipedia and realized that although he did a lot in quantum theory and was an accomplished physicist, he was largely forgotten by the physics community by 1970s. Nothing explained the aura that I remember from my ITEP days. Gennady Gorelik, a well known historian of science who specializes on Landau and his circle,

---

[44]For references see p. 109.

suggested that perhaps this aura came from Landau himself through his students. More than once in his studies Gorelik saw evidence that Landau had held Placzek in high esteem.

In the beginning of my narrative I have to explain the background. The story I will tell you below started in 1936 in Kharkov, Ukraine, at the Ukrainian Physical Technical Institute (UPTI). It spanned four years and two continents. Placzek visited UPTI twice: first, for about seven months, from May till December 1933[45] and then, three years later, for about six weeks in December 1936-January 1937, as is clear from Alexander Akhiezer's reminiscences[46] on page 77. It is likely that Placzek arrived in Kharkov in late November of 1936. In both cases his primary goal was to work with Lev Landau. The very fact of Placzek's *two* visits to UPTI is confirmed, for instance, in an official speech by I.V. Obreimov, the first UPTI Director [39].

Not much is known about his first visit, except that it resulted in the Landau-Placzek paper [40] (Figure 1.26, page 68). Another "material evidence" of his eight-month stay in Kharkov in 1933 was the arrangement regarding the Russian translation of his book on the Rayleigh Scattering which was released in Kharkov in 1935, see Figure 1.27.

When Landau moved to Kharkov as Head of the UPTI Theory Department in August 1932, the Institute was on the rise, high expectations of converting it into a first-class international physics center were shared by many enthusiasts, and foreign experts were welcome in the USSR. Apparently Placzek's stay was uneventful. So far, no mentions were found in archives or the memoir literature. His second rather short stay in Kharkov was different. It occurred in the stormy atmosphere of 1936-37 when the advent of the Great Terror was palpable. It coincided with a rather tragic chain of events at UPTI, in which Georg Placzek was right at the epicenter.

The appearance of foreign physicists in Kharkov dates back to 1931. In March of that year, Alexander Weissberg arrived in Kharkov. He became the first senior foreign physicist hired to work at UPTI on a long-term basis. Born in Poland and educated in Vienna and Berlin, Weissberg was a talented technical physicist. At that time he was an ardent communist, as one can see from his personal Comintern[47] file

---

[45] See Document 1.25 and its transcription on page 234.

[46] See also the NKVD interrogation file of Konrad Weisselberg in Chapter 2, page 125.

[47] The Communist International, abbreviated as Comintern, was an international communist organization that advocated world communism. The International intended to fight "by all available means, including armed force, for the overthrow of the international bourgeoisie and for the creation of an international Soviet republic as a transition stage to the complete abolition of the

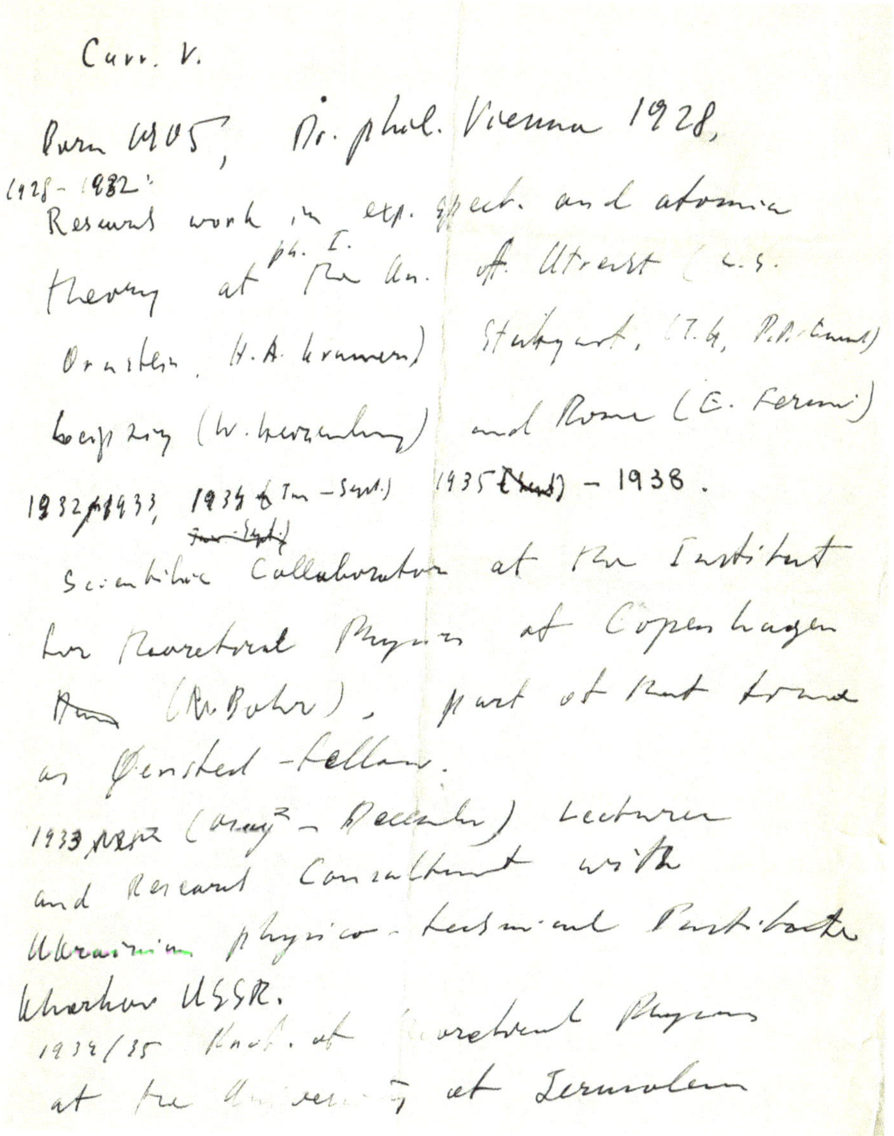

Curr. V.

Born 1905, Dr. phil. Vienna 1928,

1928–1932: Research work in exp. spect. and atomic theory at the Un. of Utrecht (L.S. Ornstein, H.A. Kramers) Stuttgart, Leipzig (W. Heisenberg) and Rome (E. Fermi)

1932/1933, 1934 (Jan.–Sept.) 1935 – 1938.

Scientific Collaborator at the Institute for Theoretical Physics at Copenhagen (N. Bohr), part of that time as Oersted-fellow.

1933 (May – December) Lecturer and Research Consultant with Ukrainian physico-technical Institute Kharkov USSR.

1934/35 Prof. of Theoretical Physics at the University of Jerusalem

Figure 1.25 The first page of Placzek's CV which he drafted (presumably) in Montreal. Undated. Courtesy of AIP.

(Figure 1.28). I mention this detail since it will be important in what follows.

Weissberg's plan was to stay in the USSR for good and to establish himself as part of a paradise state of workers and peasants. He quickly

State." Comintern's Headquarters were located in Moscow. It was dissolved by Joseph Stalin in 1943.

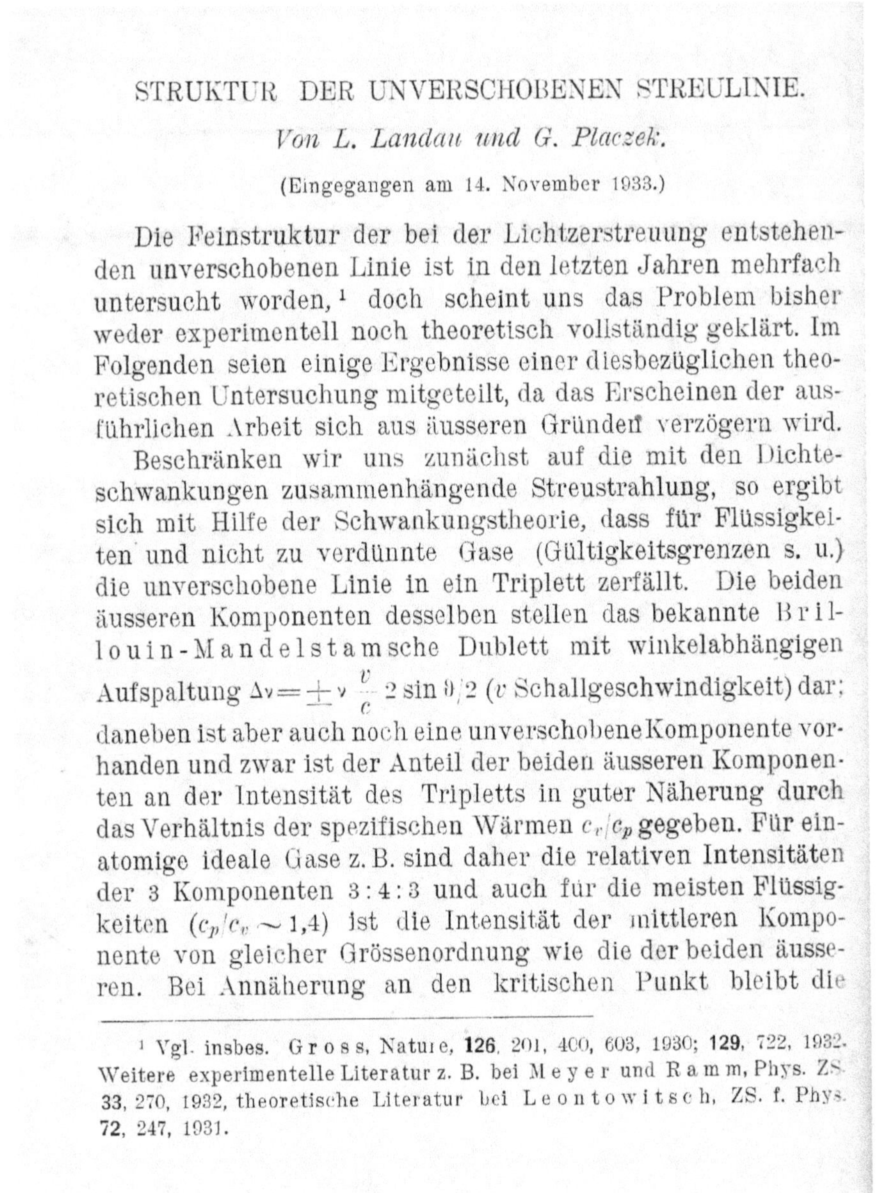

STRUKTUR DER UNVERSCHOBENEN STREULINIE.

Von L. Landau und G. Placzek.

(Eingegangen am 14. November 1933.)

Die Feinstruktur der bei der Lichtzerstreuung entstehenden unverschobenen Linie ist in den letzten Jahren mehrfach untersucht worden,[1] doch scheint uns das Problem bisher weder experimentell noch theoretisch vollständig geklärt. Im Folgenden seien einige Ergebnisse einer diesbezüglichen theoretischen Untersuchung mitgeteilt, da das Erscheinen der ausführlichen Arbeit sich aus äusseren Gründen verzögern wird.

Beschränken wir uns zunächst auf die mit den Dichteschwankungen zusammenhängende Streustrahlung, so ergibt sich mit Hilfe der Schwankungstheorie, dass für Flüssigkeiten und nicht zu verdünnte Gase (Gültigkeitsgrenzen s. u.) die unverschobene Linie in ein Triplett zerfällt. Die beiden äusseren Komponenten desselben stellen das bekannte B r i l l o u i n - M a n d e l s t a m sche Dublett mit winkelabhängigen

Aufspaltung $\Delta\nu = \pm\,\nu\,\dfrac{v}{c}\,2\sin\vartheta/2$ (v Schallgeschwindigkeit) dar; daneben ist aber auch noch eine unverschobene Komponente vorhanden und zwar ist der Anteil der beiden äusseren Komponenten an der Intensität des Tripletts in guter Näherung durch das Verhältnis der spezifischen Wärmen c_v/c_p gegeben. Für einatomige ideale Gase z. B. sind daher die relativen Intensitäten der 3 Komponenten $3:4:3$ und auch für die meisten Flüssigkeiten ($c_p/c_v \sim 1{,}4$) ist die Intensität der mittleren Komponente von gleicher Grössenordnung wie die der beiden äusseren. Bei Annäherung an den kritischen Punkt bleibt die

---

[1] Vgl. insbes. G r o s s, Nature, **126**, 201, 400, 603, 1930; **129**, 722, 1932. Weitere experimentelle Literatur z. B. bei M e y e r und R a m m, Phys. ZS. **33**, 270, 1932, theoretische Literatur bei L e o n t o w i t s c h, ZS. f. Phys. **72**, 247, 1931.

Figure 1.26　The first page of the Landau-Placzek paper received by the Editorial Office of Physikalische Zeitschrift der Sowjetunion on November 14, 1933.

made friends, who started calling him Alexander Semyonovich, following traditional Russian usage of a first name and patronymic. He knew many Western physicists, and he was supposed to entice to Kharkov the most prominent of those, who potentially might accept. In 1933

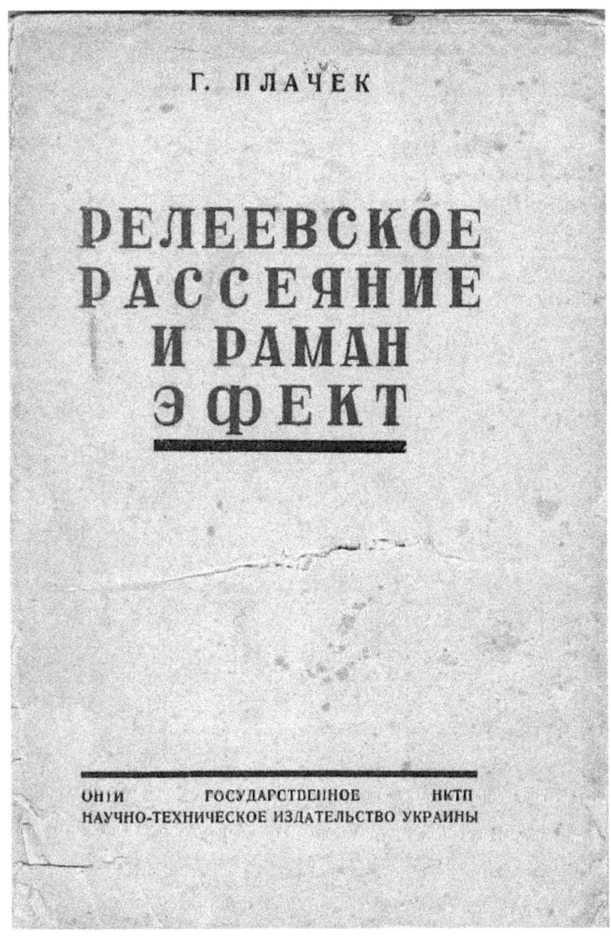

Figure 1.27   The cover page of the 1935 Russian edition of Placzek's book. Ukraine State Publishers on Science and Technology.

he was put in charge of construction at UPTI of the largest cryogenic laboratory in Europe. This assignment opened to Weissberg the doors of the highest offices in the Soviet Government.

Shortly after Weissberg's relocation to Kharkov, Éva Striker, an artist and ceramics designer, joined him as his fiancée. Éva came from an assimilated Jewish Hungarian family. Later they married.[48]

Their marriage lasted only a few years, ending in separation in 1934

[48] In 1938 after her second marriage in England, Éva became Eva Zeisel. After subsequent emigration to the US she eventually made her way to the summit of the artistic Olympus and became an internationally acclaimed designer. Eva Zeisel received many distinguished awards, e.g. the 2002 Living Legend Award from the Pratt Institute, the Middle Cross of the Order of Merit award of the Republic of Hungary (2004), the 2005 Cooper-Hewitt National Design Award for Lifetime Achievements. Eva Zeisel died in 2011 at the age of 105.

Alexander W e i s s b e r g (Peter Warbeck)   Berlin-Brandenburg
                                                      9. Verwaltungsbezirk

geb. 8. 10. 1901 in Krakau

Beruf: Physiker

soz. Herkunft: nicht ersichtlich

Staatsangehörigkeit: Österreicher

Pol. org.: SPÖ 1920 bis 1927
           KPÖ seit 1.5.1927
           KPD seit 1.12.1929

Gewerkschaftl. org. 1927 Bund der Industrieangestellten in Wien

Einreise in die SU: 21.3.1931. Er kam zur Arbeitsaufnahme mit Zu-
     stimmung der Partei auf Anraten des Parteiapparats, da ein wei-
     teres Verbleiben im Interesse der Partei nicht wünschenswert war.

Charakteristik und Parteitätigkeit:
     Von seiner Parteitätigkeit in Österreich liegen nur die Angaben
     vor, dass er in Wien im II. Bezirk der Partei angehörte. Die
     österreichische Sektion beim EKKI bestätigt am 25.7.1936 die Mit-
     gliedschaft vom 1.5.1927 bis 1.12.1929 in der KPÖ. Sie befür-
     wortet seine Überführung in die KPdSU(B) und seinen Übertritt von
     der KPÖ zur KPD. W. will bereits in den Jahren 1925/27 für die
     KPÖ fraktionell in der sozialistischen Jugend in Österreich ge-
     arbeitet haben.

     In Deutschland war er Student und gehörte der Halenseer Partei-
     organisation der KPD an. Dann war er kurze Zeit in Gelsenkirchen.
     An der allgemeinen politischen Parteiarbeit hat er nicht teilge-
     nommen. Der Parteiapparat (Edgar) wollte ihn nachrichtenmässig
     auf Grund seiner Tätigkeit in der technischen Hochschule zur
     besonderen Arbeit heranziehen. Zur praktischen Arbeit kam es
     jedoch nicht, da er auf einer Eisenbahnfahrt einem Freund er-
     zählt hat, er arbeitet für eine russische Spionagestelle. Er
     wurde darum sofort von den ihm bereits bekannten Genossen isoliert
     und man schickte ihn in die SU, da evtl. die Partei durch ihn
     kompromittiert werden konnte.

     Genosse Edgar charakterisiert ihn als für die konspirative Arbeit
     unbrauchbar, unerfahren und wenig diszipliniert. Er hält ihn für
     einen Quatschkopf.

     Der Genosse Houtermanns und der Genosse Karl Höflich kennen ihn
     seit 1929 als einen guten Genossen.

     In der SU arbeitete er in einem Charkower Institut. Von diesem
     Institut war er bereits zweimal in offizieller Mission nach Deutsch-
     land geschickt worden. Von der russischen Parteiorganisation lie-
     gen gute Charakteristiken über ihn vor.

     1936 ersuchte er in einem Brief an die österreichische Sektion
     beim EKKI um die Erlaubnis, zur Parteiarbeit nach Österreich zurück-
     fahren zu dürfen, da er dort legal leben könnte und die Möglich-
     keit habe, in einem Ingenieurbüro eines Freundes zu arbeiten.

     Die deutsche Vertretung beim EKKI ersucht er gleichfalls um die Er-
     laubnis, nach Österreich fahren zu dürfen und die evtl. Geneh-
     migung der Österreichischen Sektion zu übermitteln.

                                                            ./.

Figure 1.28   One of the pages of Weissberg's Comintern file, courtesy of RGASPI.

when she left Kharkov for Moscow. On May 26, 1936, Éva was arrested
by the NKVD.[49] One of the charges brought against her was her alleged

---

[49]The Soviet acronym for the name of the secret police at that time. The Soviet secret police has
changed acronyms like a chameleon. It started out as the Cheka, and then became the GPU, the
OGPU, the NKVD, the NKGB, the MGB, and finally the KGB. Even today, however, people often

participation in a plot to assassinate Stalin. After a few days of imprisonment in Moscow's *Butyrki*, she was transferred to a Leningrad prison where she was held for 16 months, 12 of them in solitary confinement.[50]

Three months later, Lev Kamenev and Grigory Zinoviev, two highest-ranking Bolshevik leaders, were arrested on charges of forming a terror organization with the purpose of killing Joseph Stalin. The show trial that followed was covered as front-page news in major world media. Kamenev and Zinoviev were sentenced to death on August 24, 1936, and executed by firing squad the next day.

In 1936, Georg Placzek traveled to the USSR with Victor Weisskopf who was also invited by P. Kapitza. This journey is described by Weisskopf in [38], pp. 101-102:

> I accepted an invitation from Kiev to go to Russia, to talk things over both with the Kiev people and with Kapitza. I did it because it gave me an opportunity to observe first-hand the changes that had taken place under the steadily increasing terror of Stalin, and I was eager to take Ellen on her first trip to the Soviet Union. As it happened, George Placzek was invited to the USSR at the same time, so we planned to travel together.
>
> Placzek made a list of things that were hard to get in the USSR to bring to our Soviet friends. Among many other items he decided to include a large quantity of condoms. Since he had a very generous nature, he bought about a thousand of them. We decided to assemble all the goods in our apartment. On the appointed day, Placzek arrived with his arms loaded with packages. He began to unpack the boxes of chocolate, slide rules, tissues, and soaps he had bought. Then he suddenly stopped, hit his forehead with the palm of his hand, and said, "The Copenhagen taxi trade is going to die out!" We couldn't imagine what he meant. "I forgot the box with the condoms in the taxi," he finally told us, and he left to buy another huge batch.
>
> *We left for the USSR without any illusions, but what we found was even more dispiriting than what we had*

> *expected. We were instantly and sharply aware of an atmosphere of fear and terror. Some of the friends we called acted as if they had never met us. (Among these was my old friend Yuri Rumer,[51] who was put into a camp not long after our visit.) We found that most people were afraid of contact with foreigners, except on official or professional business such as the talks we delivered at the university or in research institutes. If I had had any doubt about it, I was quickly convinced that a job in Russia, even under the most generous of terms, would have been most problematic. During that time, the winter of 1936, the fear was most apparent in Moscow...*

It is likely that Placzek intended to stay in Kharkov until a paper he had hoped to write with Landau would be finished. It never happened, however, although traces of this unfinished paper can be found in Landau's textbook [42]. Lev Landau hastily left Kharkov for Moscow around January 15, 1937, apparently anticipating an imminent *pogrom* at UPTI.

---

[51] Weisskopf writes Sasha Rumer, which is an obvious mistake. Yuri Rumer was arrested in Moscow on April 28, 1938, simultaneously with Lev Landau and Moisei Koretz.

The fluctuations of entropy have zero frequency, as stated above, and so scattering by them gives a further central line with $\Omega = 0$ (L. Landau and G. Placzek, 1933).[†]

Let us determine the intensity distribution of the undisplaced scattering between the doublet and the central line. By the intensity of the doublet we mean the sum of those of its components, i.e. twice that of either one separately.[‡] The total extinction coefficient given by (120.2) or (120.3) is $h = h_d + h_{cl}$.

Since the doublet lines are due to scattering by adiabatic pressure fluctuations, their intensity is given by the second term in (120.3), which arises from these fluctuations. The adiabatic derivative $(\partial\varepsilon/\partial\rho)_s$ can be related to the isothermal derivative by changing to the variables ρ and T:

$$\left(\frac{\partial\varepsilon}{\partial\rho}\right)_s = \left(\frac{\partial\varepsilon}{\partial\rho}\right)_T + \frac{T}{c_v\rho^2}\left(\frac{\partial P}{\partial T}\right)_\rho\left(\frac{\partial\varepsilon}{\partial T}\right)_\rho.$$

If the temperature dependence of ε at constant density is neglected, then $(\partial\varepsilon/\partial\rho)_s = (\partial\varepsilon/\partial\rho)_T$. To the same accuracy, in the total derivative written in the form (120.2) we may neglect the second term; see Problem 1 for a calculation in which these approximations are not made. Lastly, using a known thermodynamic formula for the ratio of the adiabatic and isothermal compressibilities (see *SP* 1, (16.14))

$$\left(\frac{\partial\rho}{\partial P}\right)_s = \frac{c_v}{c_p}\left(\frac{\partial\rho}{\partial P}\right)_T, \tag{120.7}$$

we obtain the *Landau–Placzek formula* for the doublet part of the total undisplaced line intensity:

$$h_d/h = c_v/c_p. \tag{120.8}$$

To determine the shape of the lines, it is necessary to consider the different-times correlation function and take account of the dissipative processes which cause the damping of the fluctuations. For the pressure fluctuations, these are viscosity and thermal conduction. The Fourier components of the correlation function for adiabatic pressure fluctuations are

$$(\delta P^2)_{\Omega\mathbf{q}} = \frac{\rho T u^3 \gamma}{(\Omega \mp qu)^2 + u^2\gamma^2}, \tag{120.9}$$

where

$$\gamma = \frac{q^2}{2\rho u}\left[\tfrac{4}{3}\eta + \zeta + \kappa\left(\frac{1}{c_v} - \frac{1}{c_p}\right)\right]; \tag{120.10}$$

see *SP* 2, §89. The quantity γ is the sound absorption coefficient per unit length; η and ζ are the viscosity coefficients and κ the thermal conductivity of the medium (see *FM*, §77). The intensity distribution in the line (in each doublet component), for a given direction of scattering, is proportional to (120.9). Normalization to unity gives

$$dI = \frac{\Gamma}{2\pi\left[(\Omega - \Omega_0)^2 + \tfrac{1}{4}\Gamma^2\right]}\,d\Omega, \tag{120.11}$$

[†] In superfluid liquid helium (the isotope ^{4}He), entropy perturbations are propagated as weakly damped vibrations called second sound waves, whose velocity u_2 is, however, much less than that of ordinary sound. The central scattering line in superfluid helium is therefore split into a narrow doublet, whose width is given by the same formula (120.6) with u_2 in place of u (V. L. Ginzburg, 1943).

[‡] The difference between the intensities of the two components is usually, according to (118.4), quite unimportant, since $\hbar\Omega_0 \ll T$.

Landau and Lifshitz's *Electrodynamics of Continuous Media*, Butterworth-Heinemann; Second Edition, 1984.

The final result is

$$\frac{h_{cl}}{h_d} = \left(\frac{c_p}{c_v} - 1\right)\left[\left(\frac{\partial \varepsilon}{\partial T}\right)_P \bigg/ \left(\frac{\partial \varepsilon}{\partial \rho}\right)_s \left(\frac{\partial \rho}{\partial T}\right)_P\right]^2.$$

In the Landau–Placzek approximation, the expression in the square brackets is unity.

PROBLEM 2. Light is scattered in a gas whose molecules are linear, with polarizabilities α_1 and $\alpha_\perp$ along and across the axis respectively. Determine the intensity resulting from the various types of scattering.

SOLUTION. The total intensity of scattered light (for given vibrational and electronic states of the molecules) includes the Rayleigh scattering and the rotational part of the Raman scattering. Since the scattering takes place independently at the individual molecules of the gas, the total extinction coefficient is most simply obtained from formula (92.4), by multiplying by the number of particles per unit volume N and replacing $|\alpha V|^2$ by $\frac{1}{3}\alpha_{ik}^2 = \frac{1}{3}(\alpha_1^2 + 2\alpha_\perp^2)$:

$$h = \frac{8\pi\omega^4 N}{9c^4}(\alpha_1^2 + 2\alpha_\perp^2); \tag{1}$$

the polarizabilities as defined here and in §92 differ by a factor V.

The undisplaced Rayleigh line is due to the scalar part of the polarizability, i.e. it is the same as if the polarizability tensor of the molecule were $\frac{1}{3}\alpha_{ll}\delta_{ik}$. The same formula, (92.4), therefore gives

$$h_{undisp} = \frac{8\pi\omega^4 N}{27c^4}(\alpha_1 + 2\alpha_\perp)^2. \tag{2}$$

The difference $h_{total} - h_{undisp}$ includes the background (scattering by anisotropy fluctuations) and the rotational Raman scattering. In order to separate the former, we must first average the polarizability tensor of the molecule with respect to rotation about some particular axis (perpendicular to the axis of the molecule). The polarizability along the axis of rotation averaged in this way is evidently $\alpha_\perp$, and that along any direction in a plane perpendicular to the axis of rotation is $\frac{1}{2}(\alpha_\perp + \alpha_1)$. In other words, a molecule rotating about a given axis is to be regarded as a particle for which the principal values of the polarizability tensor are $\alpha_\perp, \frac{1}{2}(\alpha_\perp + \alpha_1), \frac{1}{2}(\alpha_\perp + \alpha_1)$. Using these, we calculate the symmetrical tensor $\alpha_{ik} - \frac{1}{3}\alpha_{ll}\delta_{ik}$, whose trace is zero, and then a procedure similar to the derivation of formulae (1) and (2) gives

$$h_{backg} = \frac{8\pi\omega^4 N}{9c^4}\frac{(\alpha_\perp - \alpha_1)^2}{6}. \tag{3}$$

Finally, the intensity of the rotational Raman scattering is obtained by subtracting (2) and (3) from (1):

$$h_R = \frac{8\pi\omega^4 N}{9c^4}\frac{(\alpha_\perp - \alpha_1)^2}{2}.$$

§121. Critical opalescence

The isothermal compressibility $(\partial \rho / \partial P)_T$ increases without limit as the critical point is approached. The expression (120.2) for the total intensity due to Rayleigh scattering therefore increases also. This indicates a marked increase in scattering near the critical point, called *critical opalescence*.† The formula (120.2) itself is, however, inapplicable, because near the critical point the single-time correlation between the density fluctuations (and therefore the permittivity fluctuations) at different points in space extends to a distance of the order of the correlation radius r_c, which increases without limit as the critical point is approached (see *SP* 1, §§152, 153). Here, therefore, we cannot in general replace the factor $e^{-i\mathbf{q}\cdot\mathbf{r}}$ in (119.9) by unity, even when calculating the total scattered intensity (and not just when calculating its spectral fine structure).

---

† The idea that this phenomenon is due to an increase in the density fluctuations was put forward by M. Smoluchowski (1908). In relation to the van der Waals theory of the critical point (see *SP* 1, §152) it was discussed by L. S. Ornstein and F. Zernike (1914).

Landau and Lifshitz's *Electrodynamics of Continuous Media*,
Butterworth-Heinemann; Second Edition, 1984.

Here I would like to continue quoting Weisskopf:

> I have never felt any regrets about that decision [to leave the Soviet Union]. Indeed, the experiences of some of my friends in the USSR confirmed the rightness of my decision. Shortly after our visit Alex Weissberg was arrested in Kharkov. One of the reasons for his arrest may have been a dangerous joke played by Placzek during that trip. In Kharkov, Placzek was offered a job at the Physico-Technical Institute. During an evening party he said he would accept that offer if five conditions were fulfilled: first, a salary of a certain amount; then part of the salary to be paid in dollars; then every year a visit to the West; then a certain number of collaborators; and lastly, "the *khasain* must go." The word *khasain* means something like "leader" referred to Stalin.[52] Unfortunately, a faithful Communist who was at that party[53] reported this to the authorities. Placzek had to leave Russia immediately.
>
> Soon after this, Weissberg, who had invited Placzek to the Soviet Union, was arrested. He was held in detention for many years...

Alexander Weissberg was arrested by the NKVD on March 1, 1937. The story of Weissberg's arrest by the NKVD and his imprisonment is described in detail in a recently released book *Physics in a Mad World* [20] In 1940, after the Molotov-Ribbentrop pact, the NKVD handed over to the German Gestapo a group of Austrian and German communists and Jews. Weissberg was in this group for both reasons. His subsequent life is no less remarkable [20]. From approximately March of 1940 till March of 1943 he was held in the Krakow ghetto. Let us remember these dates; we will need them to understand some of the letters that will be presented below in this Chapter.

On March 4, 1937, another UPTI employee, a chemical engineer, Konrad Weisselberg was arrested. We will hear about him later too. He was executed on December 16, 1937. The leading physicists Lev Rozenkevich (August 6, November 9, 1937), Lev Shubnikov (August 6, November 10, 1937), Matvei Bronstein (August 6, 1937, February 18,

---

[52]"Khozyain" in Russian means Boss or Master and at that time was a euphomism for Stalin.
[53]Barbara Ruhemann, a very rigid Communist party member, reported this joke to the Communist Party in Kharkov with the result that shortly after, the local newspaper started to attack the foreign physicists with gradually heavier and more explicit accusations of being German spies.

1938), Vadim Gorsky (September 21, November 8, 1937), etc. followed. The first date in the parentheses above is the date of arrest, the second is the date of execution.

This was the beginning of the Great Terror at UPTI. Many more were sent to Gulag. On December 1, 1937, Friedrich (Fritz) Houtermans, a well-known Austrian-German physicist who had fled German Nazis and worked at UPTI since February 1935, was arrested too. He left behind his wife Charlotte (her friends called her Schnax) and two small children in a desperate situation. They had no means of survival. At first, Charlotte did not even have her passport which was with Fritz at the moment of his arrest and was confiscated by the NKVD [20; 43].

Only a miracle helped Charlotte and her children to escape from the USSR to Riga, Latvia, which was still an independent country until August 1940. Caught between two evils – German National Socialism and Soviet communism – Charlotte Houtermans would have probably perished if it were not for the brotherhood of physicists: Niels Bohr, Wolfgang Pauli, Albert Einstein, James Franck, Max Born, Robert Oppenheimer and many noted scientists who tried to save friends and colleagues who were in mortal danger, entrapped in simmering pre-WWII Europe. Georg Placzek was a part of this brotherhood.

During Placzek's second visit to Kharkov he developed friendly relationships with both Alexander Weissberg, his former wife Éva, and the Houtermans family. When an emergency operation to bring Charlotte Houtermans to the United States and rescue Fritz Houtermans from Stalin's prison tortures started in 1938, Placzek – in Paris at that time – joined immediately. Later, in 1940, the scope of this operation was expanded with the goal of relocating Alexander Weissberg who was entrapped in the Krakow ghetto to America. Although Poland was already occupied by Nazi Germany, the United States were not at war with Germany until December of 1941.

The first part of this operation ended in success with regards to Mrs. Houtermans and her children, while it failed in the latter, see [20; 43]. Weissberg stayed in occupied Poland until the end of the war. He miraculously survived despite the fact that his relatives and close friends perished and he himself was arrested by the Gestapo more than once. He took part in the Warsaw Ghetto Uprising (April 19-May 16, 1943) and in Warsaw Uprising (August 1-October 2, 1944).

Placzek's correspondences pertinent to the rescue effort – sometimes heartwarming, sometimes naive, and sometimes bitter – are presented below (pages 83-107).

***

To fully understand certain nuances in Placzek's letters the reader should take into account the following.

1) Rumors that Placzek caused mass arrests among Kharkov physicists started circulating among those who knew about them shortly after Placzek's return to the West, despite the absence of any solid evidence. In the 1990s many documents regarding the UPTI *pogrom* were declassified. The interrogation file of Konrad Weisselberg (see Chapter 2) shows that the NKVD was indeed interested in Placzek who is mentioned in the file more than once. At first, the NKVD insisted that Placzek had organized a German spy network, but then switched to a "legend" according to which Placzek had prepared a Trotskyite plot in Kharkov.

However, from what we know today about the mechanisms of the Great Terror, we can conclude that Placzek could have been a pretext for the arrests, but he certainly was not the reason. Even if he never showed up in Kharkov, another pretext would have been found.

2) Although it seems likely that Placzek cut short his second visit to UPTI, he could not have left Kharkov earlier than Landau.[54]

A. Akhiezer's recollections [44] seem to imply this:

> Placzek was in a very close relationship with Landau who was my thesis adviser. Moreover, when I defended my Candidate Thesis[55] in late December of 1936, Placzek was on my defense committee. All children around his dwelling knew him and always teased: "Placzek, Placzek, giv'me a *kalatchik*."[56] He was incredibly witty, one can rarely encounter such a joker. He left before other [foreigners] were deported [or arrested].

This narrative must be supplemented by Konrad Weisselberg's statement (see Weisselberg's NKVD interrogation file, page 125)·

---

[54]Landau left Kharkov around January 15, 1937.

[55]The Russian analog of Ph.D.

[56]*Kalatchik* in Russia is a small round loaf in the form of a ring with a small hole, similar to the American bagel. The part of the interview with A. Akhiezer relevant to the quotation above can be found on page 289.

In 1936 Placzek, a Czech, came to visit Weissberg. He stayed with us for about 6 weeks. There was also Victor Weisskopf, an Austrian who'd come from Copenhagen. He and his wife stayed with us for 5 days or so.

3) Placzek's political views in the early 1930s can be characterized as leftist. At that time, probably in response to the growing menace from the Nazi ideology spreading from Germany all over Europe, the majority of the European *intelligentsia* shifted to the left wing of the political spectrum. Many became communist sympathizers; some, like A. Weissberg and F. Houtermans, joined the ranks of the communist parties associated with Comintern.

Edward Teller testifies [45]:

> Returning to the university that fall, I set about my research with renewed diligence. I had met another young physicist, George Placzek. Czechoslovakian by birth, Placzek had moved to Vienna before coming to Göttingen. He was only a few years older than I, but he had a much greater knowledge of physics. He reminded me of my friend Landau in two other ways: He was also a communist, and he was wickedly sarcastic about the stupidities of the social system in Central Europe.
>
> I had been interested in communism since my original negative experience with it as an eleven-year-old in Hungary, but my childhood opinion had been considerably modified during the seven years I spent in Germany between 1925 and 1933. Several close associates presented me with the ideas and ideals of communism. Eric Landauer, a dear friend of my Karlsruhe mentor Feri von Körösy, was the first dedicated communist I knew; and two of my later colleagues, Lev Landau and George Placzek, were also convinced exponents of communism....
>
> But I was also impressed by Geo Gamow's deep resentment of communism.

Placzek was never a card-carrying member of any party. Moreover, he was among the first to realize that extreme left and extreme right represent in fact one and the same cause – not in words, of course, but in methods, deeds and goals. The events he witnessed in 1936 in Kharkov, the atmosphere of fear and terror, served as an eyeopener. That was Placzek's moment of truth. He returned to the West with a clear-cut understanding that communism of the Stalin type is a dictatorship in no way better that Hitler's national socialism. I would say – although I cannot directly prove it – that Placzek's insight into the communist idealism and dogmas was even deeper. Perhaps, he even foresaw the truth of the Gulag (which was open to the public in 1956, after Placzek's death) and the postwar tragedies in East Europe.

Alexander Weissberg came to similar conclusions in the late 1940s. In 1950 he joined the Congress for Cultural Freedom, an anti-communist advocacy group founded in that same year.

As for Fritz Houtermans, after the end of the war he never resumed his membership in the German Communist Party. However, he did not give up his leftist sympathies (deciding to keep his political persuasions to himself, though, see [20]).

Utopian socialist ideas originated in Europe. In the 1930s the European *intelligentsia* was infatuated with communism, for obvious reasons. Even such a great German writer as Lion Feuchtwanger who was vigorous in exposing the brutality of the Nazis, failed to see the brutality of the rule of Joseph Stalin and, moreover, praised him. Romain Rolland, a distinguished French intellectual and writer, was an unconditional apologist for Stalinism – like many, he was blinded by the "sacred idea."

The American *intelligentsia* by and large looked at the ideas of communism with much less enthusiasm. The decision-making echelons were suspicious of communists and their sympathizers, to say the least, associating them – in many instances, rightly so – with the Comintern and orders from Moscow.[57] The latter remark provides a better perspective on the contents of letters on pages 97-105 which clearly exhibit Placzek's reluctance to reveal his sponsorship of Alexander Weissberg. He was by far not the only one. Felix Bloch, the Nobel laureate by then settled at Stanford, was at pains to explain to Eva how his American colleagues, upon learning that Houtermans and Weissberg were Communist Party members, refrained from assistance [47]. I find Felix Bloch's letter to Eva very instructive in this respect. Although it is not directly related to Placzek, I decided to reproduce it in its entirety at the end of the corresponding section (page 106).

---

[57]It is no secret now that practically all spies working for the Russian intelligence within the Manhattan Project were communists, see e.g. [46].

***

I hope that the brief introduction provided above helps to set the scene: pre-war and war-torn Europe and two physicists in a desperate situation: Fritz Houtermans in an unknown NKVD prison somewhere in the USSR, and Alexander Weissberg given by the NKVD into the hands of the Gestapo which in turn confined him to the Krakow ghetto destined for extermination. Two deeply involved women – Charlotte Houtermans and Eva Zeisel – and Georg Placzek among others try to do their best to save them. This human drama is narrated in thirteen old letters written by George Placzek or addressed to him which the reader will find below.

Figure 1.29 Left to right: Robert Millikan, Jean Baptiste Perrin, George Placzek, Guido Beck, Samuel Goudsmit, Walter Heitler, Arthur Compton and Giovanni Giorgi in Rome, 1931. Courtesy of AIP Emilio Segrè Visual Archives, Goudsmit Collection.

Figure 1.30   Niels Bohr and George Placzek at NBI, 1935. Credit: Niels Bohr Archive, Copenhagen.

Letters

Placzek to Charlotte Houtermans [58]

April 12, 1938

Dear Schnax:

Thank you very much for the regards sent to me via Møller. You have not heard anything from me so far because a rumor spread by Adams [59] after my visit in London says that Fisl [60] is free in Kharkov, so further messages about this were awaited and actions in this matter came to a stop. As we have learned from Peierls [61] and as it probably became clear from your letter to Møller as well, this was a misunderstanding.

Now about the two current questions.

1) Franck wrote to Bohr a while ago about Einstein's and his intervention with Troyanovsky. [62] Nevertheless, it certainly would not hurt if you wrote to Franck yourself as well. The value of any intervention in Russia on Einstein's part, at any rate, appears highly doubtful.

2) Considering Bohr's letter to Russia mentioned by Mrs. Bohr, he has developed doubts about the expedience of such a step and has not sent the letter yet.

We will keep you up to date about whatever might happen further.

At any rate, there is an adequate solution to the main problem that was troubling you in London, which is obtaining all the papers and visas that will be necessary for F. after his release.

There is nothing new so far. Møller has gone on Easter vacation and sends you many kind regards. Mrs. Møller is well again. Please let us hear from you soon, one of us will surely overcome our common laziness in writing then.

Many kind regards and all the best wishes,

---

[58] See Figure 1.31 on page 85. If not stated to the contrary, the letters presented in this chapter are courtesy of the Houtermans family.

[59] Walter Adams was the General Secretary of the London-based Academic Assistance Council (AAC) which later became known as the Society for the Protection of Science and Learning. Currently, the Council for At-Risk Academics.

[60] Fritz Houtermans' nickname.

[61] Sir Rudolf Ernst Peierls, (1907-1995) was a German physicist who played a major role in Britain's nuclear program. He is known for many outstanding contributions in quantum field theory and nuclear physics.

[62] Alexander Troyanovsky (1882-1955), was the Soviet Ambassador in Washington in 1933-1938.

Yours G. Placzek

P.S. Best regards from Mr. R. Peierls too. [in handwriting - translator's note]

∽

Georg Placzek to Charlotte Houtermans

Copenhagen
April 19, 1938

Dear Schnax:

Thank you very much for your last letter that Møller and Rosenfeld answered for me. Joliot has been here now and Bohr and I have talked with him in great detail. He has already started some preparatory actions and wishes to wait for their results before he contemplates writing a letter to S[talin]. Thus Bohr will wait, too. It will probably take about two weeks. Then Bohr will meet Langevin and there will probably be a basis for further decisions.

What Joliot would like to receive from you are all the <u>official</u> data about this case, i.e., the dates of his[63] arrest and all the other official messages from or about F[ritz] (the most important are his messages to Mrs. Cohn-Vossen[64]). This is necessary for Joliot so that he can show [unclear] during negotiations and does not have to let himself be intimidated by false claims like the one that F[ritz] has never been arrested at all. So please write him, the address is as following: Prof. F. Joliot, Laboratoire de Chimie Nucléaire, Collège de France, Place Marcellin Berthelot, Paris 5e.

My America trip has to be postponed over and over again because the work with Bohr just does not get finished. I have had to exchange the tickets two times so far. This time, however, things are really getting serious and I am leaving Copenhagen on April 27. So please

---

[63]Fritz Houtermans' arrest.

[64]Margot Maria Elfriede Cohn-Vossen, née Ranft (known to her friends as Friedel) was a widow of the famous mathematician Stephan Cohn-Vossen (1902-1936) who had collaborated with David Hilbert on the 1932 book *Anschauliche Geometrie*. Stephan Cohn-Vossen was barred from lecturing in Germany in 1933 under Nazi racial legislation. In 1934 he emigrated to the USSR. After Stephan's death, Mrs. Cohn-Vossen stayed in Moscow and in 1938 married Alfred Kurella (1895-1975), a German communist functionary in exile. She died in 1957 in Soviet Georgia.

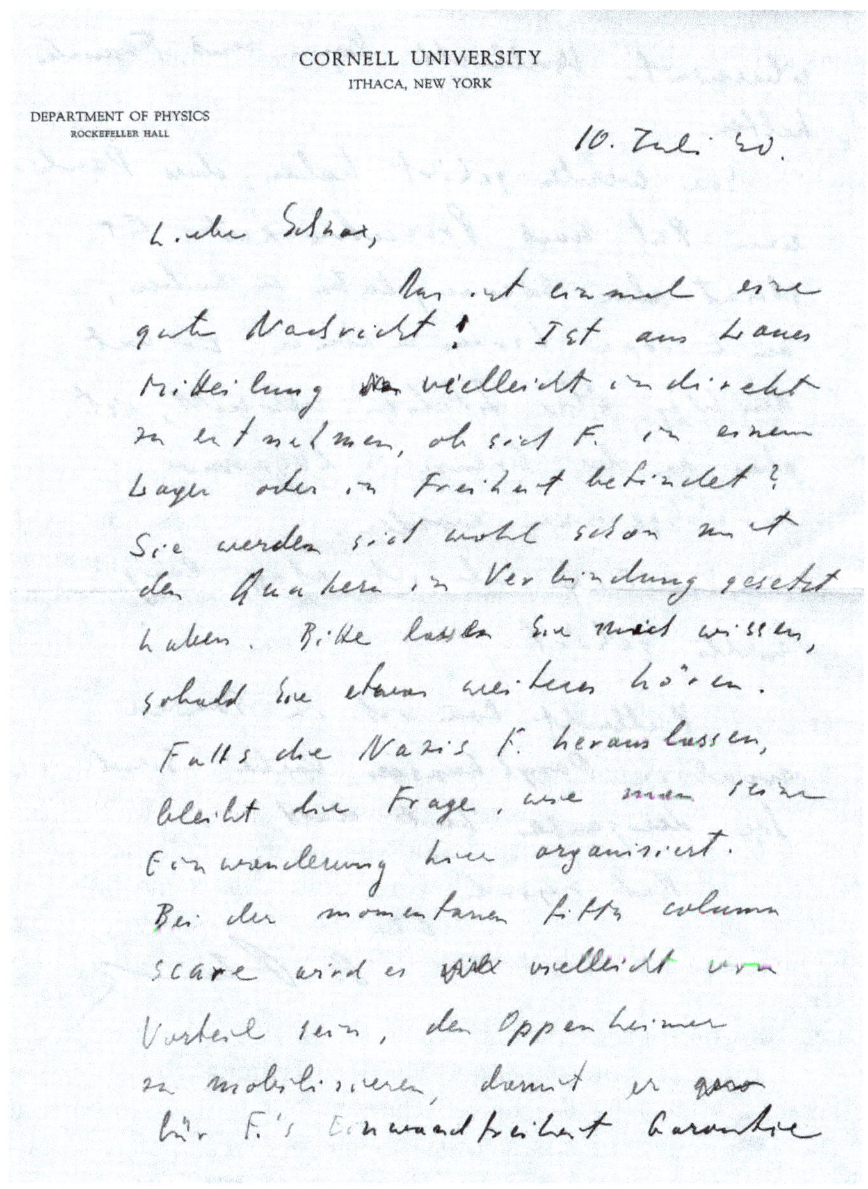

Figure 1.31   The first page of one of Placzek's letters to Charlotte Houtermans.

write me all your requests for this [illegible] before this date, as well as what is new, what plans you have etc.

Schein,[65] who is surely known to you, has finally found a job in

---

[65]Marcel Schein (1902-1960) was a Slovakian-born physicist whose research was in cosmic rays physics. Schein studied at the universities of Vienna, Würzburg and Zurich, where he got his PhD.

America (Chicago) after a long and agonizing time.

What do you think about Ruhemann's letter[66] published in *Nature* about the standard of living of the Russian scientists? I think this is the *height of shamelessness!*

Many kind regards,

Yours G. Placzek

∽

George Placzek to Eva Zeisel[67]

University Institute for Theoretical Physics
Blegdamsvej 15,
Copenhagen,
Ø[sterbro district]
May 15, 1938

Dear Eva:

Thank you very much for the letter collection. I have waited for a couple of days before I answered because I wanted to wait until the discussion of these things with Joliot and Bohr. The latter has happened now. The result is the following: Joliot has some preparatory activities running and wishes to wait for their results before he makes up his mind about writing to S[talin]. This will probably be in about two weeks.

Langevin would participate in this letter as well. I have worked very hard for the matters of Alex's and the Houtermans' to be dealt with together and hope I have achieved that. Joliot finds it important to have the data about the course of these matters so as not to show any weaknesses if there are questions. Please write him (Prof. F. Joliot, Laboratoire de Chimie Nucléaire, Collège de France, Place Marcellin Berthelot [sic], Paris 5e) an account with the exact date of his arrest

---

He taught at several European universities until 1938 when he emigrated to the United States. At the University of Chicago he became a member of staff at the Institute of Physics and was promoted to full professor in 1946.

[66]Ruhemann's letter "Salaries in Soviet Universities," was published in *Nature*, vol. 141, p. 792 (1938). For more details see *Physics in a Mad World*, Ed. M. Shifman (World Scientific, Singapore, 2015), pp. 20-28 and 138-140.

[67]Translated by Alexander Tschernow.

and the data of all the official messages later, such as the notification for the Embassy. In this letter, refer to me.

Further. Since I do not know Einstein personally, it would be pointless for me to write him. However, I am fairly close to the point of going to America and will try to get in touch with him there. Considering the Millikan action, I have written to Bloch just like you had suggested. He will see to the rest. We will keep you up to date about the coming events. I will do this from America and from here, a friend of mine will do this – Dr. Christian Møller, Institut for teoretisk Fysik, Bledgamsvej 15, Copenhagen. I would like to ask you to give him your current address and to inform him if you learn something new about Alex's matters.

I am very pleased that you still had time to leave Austria. Considering further, you should try everything to be able to stay in England, shouldn't you? Here in Denmark, as small a country as it is, the situation is very difficult, and it is virtually impossible for new emigrants to receive a residence or work permit. I could not yet find out whether there is a special journal for ceramics. At any rate, I will write you on this topic again. I would have liked to meet you, but there are no possibilities any more. Early next week, I will be in Paris (address: Hotel Trianon-Palace, 3 rue de Vaugirard, Paris 5e) and will depart on May 25.

Please write to our Paris address at any case, I will write you my American address later.

Many kind regards and all the best,

Yours
G. Placzek

P.S. If you do not really feel like writing directly to Joliot, you can send to my Paris address as well. However, for many different reasons, I consider you writing him directly the better idea. Send my regards to Prof. Polanyi.

Placczek to Eva Zeisel [68]

Paris
June 1, 1938

Dear Eva:

Thank you very much for your letter. My departure has been delayed by a few days, as you probably have learned from Eva. To the points:

1) To my utmost regret I had found out that the data considering Alex that I asked you to send to Joliot (in my last letter) did not reach him. This is a great pity! Did you not write the letter, forget to send it or did it get lost? At any rate, would you please write to Joliot immediately. I will repeat it: a) the date of Alex's arrest and b) any official news since this day are needed. This is very important. Joliot is willing to write to S[talin] considering [not really legible, probably just typing mistakes] Alex's matters. Your data are all he needs.

2) Considering your news about Alex – they are, of course, very sad, but not surprising in the least, neither to Joliot nor to me, since we both know about this method and the fact of its use in other cases. For further steps, we cannot use what you wrote us in its current form. Would you please immediately write a memorandum containing your informant's exact statement and the way you got it, then send copies of it to Joliot and to Bohr (the latter via Møller), marking them "TRES CONFIDENTIEL (VERY CONFIDENTIAL)."

I vouch for the absolute discretion. The information you have given us in the form you have used is, unfortunately, completely useless, not because we do not believe it (we do), but because it cannot be used to base anything on it.

3) I did not act upon your suggestion to have Koestler contact Joliot and intervene because I know Joliot better than Koestler does, so I did it directly. It is important that Joliot gets your data immediately. Please send him your data right away and the memorandum mentioned in 2) one day later, since you need some peace and [?] to formulate it [?] [page probably missing, since the next page does not contain a proper continuation of the phrase]

[...]

$4\frac{1}{2}$) As to Einstein, it is thought that an intervention on his part in

---

[68]Translated by Alexander Tschernow.

Russia would do more harm than good. Please discuss this with Prof. Peierls.

5) I will pursue the Millikan action further in America.

6) I hope to meet your brother in New York. Considering Denmark, I have sought competent counsel just to make everything sure. The non-surprising result is that the possibilities there are nil. I did not manage to find out the address of the special journal for ceramics. If you are still interested in it, please write to Biblioteksafdeling, Kunstindustrimuseet, København.

My American address is in the letter to Schnax. I had Peierls bring you the amber necklace Schnax had given me for safekeeping.

All the best,

Yours,

G. Placzek

∽

Georg Placzek to Charlotte Houtermans

Paris

June 11, 1938

Dear Schnax:

I have received your letter while still in Copenhagen and informed Bohr and Møller about its contents. I think you would have talked with Peierls by now. Bohr is in Warsaw and seeing Langevin now, a decision will be taken soon. I have discussed the whole thing in great detail with Joliot yesterday and he is willing to write to S[talin] about F[ritz] and Alex.[69] He considered the precise information you had given highly satisfactory. Unfortunately, the parallel information about Alex from Éva[70] lacked! His other actions are continued. You will hear from

---

[69] Alexander Weissberg.

[70] Éva Striker (also spelled as Eva Stricker; 1900-2011) was Alexander Weissberg's ex-wife. In 1938, after her second marriage in England, Éva Striker became Eva Zeisel. After subsequent emigration to the US she eventually made her way to the summit of the artistic Olympus and became an internationally acclaimed designer. Eva Zeisel received many distinguished awards, e.g. the 2002 Living Legend Award from the Pratt Institute, the Middle Cross of the Order of Merit award of the Republic of Hungary (2004), the 2005 Cooper-Hewitt National Design Award for Lifetime Achievements.

Møller via Warsaw.

I will see what has to be done in America now. My address is temporarily identical with that of Weisskopf's: c/o Weisskopf, 66 Summit Drive, Rochester (N.Y.), U.S.A. Please continue using this address while writing me.

Many kind regards and all the best wishes for the meantime,
Yours

G. Placzek

P.S. A letter for Éva is enclosed.

∞

Georg Placzek to Charlotte Houtermans

Collège de France
Laboratoire de Chimie Nucléaire
Place Marcelin-Berthelot
PARIS (Ve)
Tel: ODÉON 81-60
Paris
December 8, 1938

Dear Schnax:

Thank you very much for your letter, or rather two letters, since I still owe you a response to the letter you sent to me in America. The thing is, I thought Franck had written you that we have met in Stanford and discussed the thing in detail. The letter to S[talin] was written by Franck in cooperation with Weisskopf and me. Joliot, by the way, did not send it yet because he found it was too early to do so.

Considering the information you have given me recently, I cannot draw much of a conclusion from it, either from the positive or from the negative side. Why don't you write to Joliot [71] about it? If he has anything in his hands, it will be easier for me to talk with him about it and we will have something to remind him of this, too.

In America I had a very detailed talk with Robert Oppenheimer, a person who you are probably in contact with as well. Furthermore, I

---

[71] And mention that you have written to me, too... - G.P.

could interest Urey [72] in the thing and he is willing to participate in the action. Then September [73] came and not much was possible any more. I will talk with him again when I come back to America, probably in February.

Where is Éva? The action of [Robert] Millikan [74] has ended really badly because Éva did not inform [Felix] Bloch [75] about Alex's [Communist] party membership, so the authorities had the totally wrong prerequisites for their search. Then the Russians have asked whether the authorities had knowledge about Alex possibly being in the [Communist] party. The latter reacted by great bewilderment and anger and lost interest in the cause.

You do not write how you are personally. Please feel free to make up for it and to write me about your further plans as well.

Many kind regards,

Yours

G. Placzek

∽

---

[72] Harold Urey (1893-1981) was an American physical chemist whose pioneering work on isotopes earned him the Nobel Prize in Chemistry in 1934 for the discovery of deuterium. He also played a significant role in the development of the atom bomb.

[73] The Second World War began on September 1, 1939, with the invasion of Poland by Germany.

[74] Robert A. Millikan (1868-1953) was an American experimental physicist honored with the Nobel Prize for Physics in 1923 for his measurement of the elementary electronic charge and for his work on the photoelectric effect.

[75] Felix Bloch (1905-1983) was a Swiss physicist, working mainly in the U.S. He and Edward Mills Purcell were awarded the 1952 Nobel Prize for "their development of new ways and methods for nuclear magnetic precision measurements."

Georg Placzek to Charlotte Houtermans

[coat of arms]
French Line
S. S. Normandie
January 28, 1939

Dear Schnax:

Your regards have been given in Copenhagen as you had wished. There was a great turmoil because Bohr was departing to America. I have received my visa [illegible] and will have to go by myself. Address: Physics Department, Cornell University, Ithaca, N. Y.

Have you written to Joliot? He has not told me anything.

Has Oppenheimer given a sign of life? Please write to me at Cornell how things are, when you will have time etc.

Many kind regards,

Yours G. Placzek

∽

Georg Placzek to Charlotte Houtermans

Cornell University
Ithaca, New York
Department of Physics
Rockefeller Hall
April 21, 1939

Dear Schnax:

Thank you very much for your letter that I have found now after coming back to New York from a journey. I am very glad that you have reached this continent alive and hope you will be well here and find a job soon. Should I hear of such a possibility myself, I will of course let you know as soon as possible. Cornell is quite a distance away from the big world, though – but you never know.

I have not seen B[illegible]. Szilard[76] (Kings Crown Hotel, 420W 116th Street, New York) can give you information about his whereabouts. Bohr only will stay with us for another couple of weeks, here is his address: Institute of Advanced Studies, Fine Hall, Princeton, N. J.

Let me know if you have the possibility to go to New York. Perhaps we can meet there soon.

Best regards,

Yours G. Placzek

P.S.: [illegible] St. sends her regards. She continues vegetating here in Paris.

∞

---

[76]Leo Szilard (1898-1964) was a Hungarian physicist who conceived the nuclear chain reaction in 1933, patented the idea of a nuclear reactor with Enrico Fermi, and in late 1939 wrote the letter for Albert Einstein's signature that resulted in the Manhattan Project that built the atomic bomb.

Figure 1.32   George Placzek, Richard Courant, Niels Bohr, Frederik Lunning, and Kay Bojesen, 1939 New York World Fair. May 1939. The date can be established almost exactly. Niels Bohr arrived in New York in mid-January 1939 and stayed till mid-May. George Placzek arrived in early February. The New York World Fair was open on April 30, 1939. Credit: Niels Bohr Archive, Copenhagen.

Figure 1.33   Circa 1948-1949, Princeton, NJ. Front row (right to left): George Placzek, Cécile DeWitt-Morette, Klára Dán von Neumann (tentatively), and Els Placzek (tentatively) Behind Els Placzek is Jenny Van Hove (tentatively). Standing in the second row (right to left): John von Neumann, Léon Van Hove, unknown. From Cécile DeWitt-Morette's personal photo albums, courtesy of Chris DeWitt.

Figure 1.34   Circa 1948-1949, Princeton, NJ. John von Neumann (left) conversing with George Placzek. From Cécile DeWitt-Morette's personal photo albums, courtesy of Chris DeWitt.

Eva Zeisel to Placzek[77]

New York
May 21, 1940

Dear Placzek:

Viki has probably told you how exactly we hope to set Alex free. Viki has sent the following declaration to Misi and I think you will do the same, since any (moral) risks for you are eliminated. The real point is that Cuthbert Daniels should be completely financially secure when he will empty his securities deposit account for Alex. Daniels is helping out because, on the one hand, he understands your position very well; on the other hand, he wants to help Alex, so there is not the least danger that you could be linked with the matter, whatever may happen or not happen. Because, by the way, Daniels is an assistant of Paul Lazarsfeld's, the situation is doubly secure. Viki knows this all, no matter what, and I assume you are willing to take the same "risk" as Viki, who is father of a family. This is Viki's letter:

Dear Misi: I have heard that it is necessary to guarantee some funds in order to bring Alexander Weissberg to this country. I am willing to oblige myself to put at your disposal up to $30 a month for a period of two years in case these funds are used for the direct or indirect support of Alexander Weissberg and that you are made responsible for these payments by the guarantor of these funds. Very sincerely.

Viki has requested Misi to look after this declaration as a trustee, however, he has authorized him to show them to Daniels. I ask you to send this letter to Misi immediately via special delivery so that we finally can start acting.

I have received a postcard from Alex today where he asks for your and Viki's address. I think it will be better not give it to him to save you any kind of mutual trouble.

I am glad that a solution satisfactory for everybody could be found at the end and just would like to ask you to hurry once again.

---

[77]Translated by Alexander Tschernow.   A comment by Judith Szapor (May 3, 2017): "Georg Placzek was very close with Eva Zeisel perhaps because her second husband, Hans Zeisel, was also of Czech-Jewish background. Eva wrote to him, asking to provide an affidavit for Alex Weissberg, should he manage to get out of Poland. She told me that Placzek refused, and after that their friendship was never the same." See, however, Placzek's letter on page 105.

Eva

ᴄᴐ

Eva Zeisel to Charlotte Houtermans[78]

New York
Undated
Presumably July 1940

My dear Schnaxi:

Of course, I was very pleased about your postcard and would like to know whether you are already in direct contact with Fisl. What did Bamsi say about the news? How did she react to the fact that something that used not to be true is true now? You seem to have been right in your feeling that Fisl was back in M[oscow] then: Alex wrote that all the German citizens were gathered there before being sent to Germany. I mean those who were arrested.

Yesterday, a letter from Alex came where he writes that Fisl is in Berlin and can be contacted via Westphal.[79] Is this the thing you have written about? It is not clear from his letter whether this is an answer to our telegram or [to?] a new message he has received. Will Konrad turn up, too? The husband of the girl Alex has put up in his aunt's place has arrived as well. It seems that further transports are following on. I remember that you had registered Fisl for the quota system back then. This was enviably wise, but it would hardly help Alex, since he is born in Poland and not registered anyway.

I am writing you today not only to tell you the phrase about Fisl, I also want to write you about what is happening to save Alex, since you have asked about it. I have got the horrible feeling that just so little can be done. More precisely, I have got a feeling of great anger and I think you will share it.

Alex's situation is the following: they expect him to prove he has a real possibility to leave the country – which means, a visa to a neutral country – as soon as possible. It is true, there are no traveling possibili-

---

[78]Translated by Alexander Tschernow. Some sentences of the original are not quite as organized and clear as in the translation due to strong emotions -translator's note.

[79]Seemingly there was a person called Westphal, similar, but not identical to the region's name -translator's note.

ties at the moment except via Russia... But the problem is that he just has to show this visa to the authorities. He has asked us to hurry up via different possibilities (it is not possible to send a telegram from there) – he will obviously be sent back to concentration camp soon if he cannot prove that he will be able to leave the country as soon as there will be transportation possibilities again. His messages about this were sent at a time where Italy already was at war, so the traveling possibilities were exactly as they are today.

I think that Alex does not exaggerate and that it is our duty (of Placzek and mine) to help him by any means.

Besides the phrase from Alex's letter that I wrote you, there is some more: "give my regards to Placzek, who I often think about with affection despite the havoc he has wrought." Please understand that this means just that – he has started the trouble. I know this technique from many other cases, they issue an arrest warrant based on a conversation passed on confidentially, those who listened are blamed as well – or the work of a person is investigated. But Alex's work only was investigated after his arrest. Then came this period, very long for him, when he had plenty of time to think about things – unfortunately, we know from two sources that he has been tortured, probably he thought he would be shot soon. As he told us in his last message, he surely has taken a good look at his feelings towards Placzek, probably he has thoroughly forgiven him and thought everything about him that you could think about those who have done you the worst when you believe that it is time to formulate your last thoughts.

You can take this as sarcasm or seriously just as you wish. But surely it is credible that it is really hard for Placzek to bear the consciousness of such guilt. However, even without any guilt, you can imagine what Alex would have done for any one of his friends. He has shown it on my example. He is far from perfectly faultless, as we know well, but he would have little thought about himself if a friend had been in danger. Anyway, nevermind [sic] that I am writing about this in such detail, I am really furious.

So, now about what has happened to Alex. I met Placzek about two days after I had learned that Alex is in concentration camp in Poland. At first, Placzek said that Alex had already been given a life sentence for high treason (?) in Germany, so it would not be worth it to do anything anyway. The argument went on for half an hour until I had asked him to stop (I had started to cry with fury, and this was in the coffee house where he was eating breakfast). Then I said Einstein should be asked to intervene (I do not remember in which matter exactly) – Placzek said it could not be done and he could not do it because Einstein was very busy and would consider it contradictory to the interests of physics if he was disturbed. Besides, so said he, Alex was an engineer, not a physicist. Then I asked whether a job as a university or college lecturer was within the realms of possibility – his answer was laughing at me as a dreamer. This was all we could do in our first meeting.

A few days later, Hans and a friend of his received a promise from the St. Lawrence University in Canton that they would invite Alex to a teaching fellowship if funds are provided, just in case he will not be able to come. This is kind of a financial guarantee that would be returned once Alex has reached the country. The university has invited somebody from Vienna for a teaching fellowship last year and would like to repeat it. The president said six hundred dollars a year had been enough for the first invited person, we said one thousand two hundred would be better. At that moment we thought that a written guarantee from me, Placzek, Viki (all three of them will have teacher salaries, although in the modest range of those, next year), Mischi (who has a lawyer's license) and Hans should be enough. Misi, Hans and I contacted Viki and Placzek. Misi was willing to provide the counter guarantee for both of them, that is, to hold them harmless should any liability result from this situation. From now on, I cannot maintain the chronological order any more. Viki and Placzek came, phoned and wrote one after another. They said they would lose their livelihoods should they identify with Alex's entry into the USA. They deemed it unacceptable that their names could became known in this context in one way or another. Allright [sic]. We found a nice man, a colleague of Hans', who was willing to provide a guarantee for the whole sum if we were to provide a counter guarantee. A lot of debate in Ithaca and Rochester followed, first yes, then no. Viki was willing to write Misi a letter which he was not supposed to give anybody else where he would write that he supported some guarantee for Misi. Misi would be able to sign it for the person providing the guarantee. Then came Placzek. After much to and fro conversation, we learned that there is no way he could have his name used to do anything for Alex. Viki not only demanded our assurance that the university president knows "everything" about Alex, he also visited Hans' friend, who is preparing the contacts to the university, without our knowledge, in order to control what we would say. Everything we were supposed to say was that Alex is a communist (which is not true in the first place). The end of the long to and fro debates was that all the promises were taken back and we had lost many weeks. By now, the university president has resigned, but back then, two months ago, he promised me to follow through with all this anyway if we find the money, because he would stay there another two or three months. The negotiations with Alex's two friends were very unnerving because Placzek was not willing to do anything in any way at all, led us up and down the garden path for weeks and put forth numerous arguments about himself for hours... This is not explained in real detail.

In the meantime, one fact has changed: While earlier we could say and believe with clear conscience that in the foreseeable future we will know whether Alex will get the visa and be able to come, thus, that the money will not be bounded too far into the future, now it seems like he could not travel – at least not in the moment – and we must be prepared for the sum remaining bounded for the whole year, as long as the university's invitation remains valid. This is one consequence of not providing at least six hundred dollars to the university very soon. If I do not, I doubt we will have such a possibility ever again.

The only other small (possible) chance today are the Philippines. What you need is patronage (perhaps Prof. Einstein's patronage any-way) and a deposit of 600 dollars as well. This is urgent, too, for we already had many other possibilities disappear before our very noses due to lack of money.

The present situation is that Fritz Lederer has promised 75 dollars at Misi's request. Viki will give a bit as well, we can perhaps find two hundred dollars, but the rest depends on Placzek. End of story. There is no way Placzek could be unable to borrow two hundred dollars at once, given his car and his teacher salary next year. If he cannot, nothing can be done for Alex and he will go back to concentration camp – but he can. We have squeezed money out of all of our friends for the sake of Hans' mother's visa and have alarmingly big debts. We pay back, for the time being, sixty dollars a month from what we borrowed from Hans' mother and mine. We sent money to Hans' father until he died a month ago. I have pawned my silver for the sake of my father's ship ticket, Hans' sister has sold and pawned everything she had for the sake of Hans' mother's visa and at the day my daughter will be born, I will have to pay the physician from non-existent funds. So we cannot do much more than try to get hold of the two hundred dollars. Misi and I will see that the money is managed correctly, it will probably be entirely free in a year.

Now I have told you the story. Perhaps you have an idea what to do. Perhaps I am particularly unsuited to influence Placzek, who is not at all embarrassed about his foolish sabotage of the matter in front of me. Now, after I have vented my anger about the fact that my theories about Placzek starting all this trouble is true, I will write to Viki and Placzek. After they both have confirmed that they do not want either to use their names or to give any other help for Alex's entry into the USA, which means...

Eva

∽

Eva Zeisel to Victor Weisskopf[80]

New York
Undated, before July 12, 1940

Victor Weisskopf, Cornell

    Dear Viki,

Now I want you to explain to Placzek that Alex will suffer all the same again without his help and that we must prevent it by any means. This is within the realms of possibility – but not without his (and your) help. About the six hundred dollars. We hope we can scrounge three hundred of them from different people. We have just gotten hold of a visa for Hans' mother and are deep in debts, so that we have real concerns. We have to pay back far more than one hundred dollars of debts every month and support relatives. Thus, you have to understand that we really cannot help Alex without your help. For the time being, I am concerned about paying the physician and buying the baby carriage.

Placzek and you, you have to find three hundred dollars out of the six hundred dollars if you want to help Alex. Misi told me that, at some point, you have agreed to pay 75 dollars and to pay the 75 dollars needed from Placzek as well in case he could not pay his part. Now I think it would be excellent if you could provide one hundred dollars and Placzek

---

[80]Translated by Alexander Tschernow.

the other two hundred. If Placzek does not have the money right now, one possibility is to borrow it at the Hebrew Loan Association (as we did). They lend you up to 500 dollars and expect them back in the course of nine months. The other possibility is to borrow the money from a bank using his car and his teacher's salary as guarantees that it will be paid back, but these are just suggestions. It cannot be difficult for him if he understands that a life depends on it and how much it is his duty to provide the money.

I think it is necessary that you let us, Misi and me – or a spokesperson like, say, Szilard or Schnax – organize this matter. This would help avoid any kind of red tape and make it possible to act as quickly as necessary. By the way, I do not insist on anything and let you do it as you wish. I will send the following letter to Placzek as well and I really ask you to contact him immediately to convince him that he just must help Alex, that his failure to help Alex would send the latter to his destruction – and this time it would not be due just to imprudence. It would be nice on your part, Viki, to write a short message for Alex.

Please Viki, answer me immediately. This whole thing is very painful to me, and I am feeling just so powerless to help, while Alex is counting on all of his friends. I am sure he would have helped any one of us, committing all of his person to that – and we have not done anything for him yet. Please Viki, write to Misi or me, as you wish, but hurry up.

Eva

∽

Eva Zeisel to Placzek [81]

New York
July 12, 1940

Dear Placzek:

I have received a letter from Alex yesterday. He begs for a telegram about his visa's processing status. The authorities demand that he prove he has the possibility to leave the country. (Of course he has –

---

[81] Translated by Alexander Tschernow.

as soon as it will be possible.) This is nothing new. So we still have not done anything for him...

There is one last chance to help Alex: a deposit of 600 dollars. If we do not take this chance [handwritten addition: soon], we will have to accept that it is our fault if he goes to concentration camp after all the terrible things he has experienced.

The argument that it is not worthwhile doing anything because traveling is impossible is not really valid. Firstly, Alex knows that he cannot travel right now and still begs for a visa in a neutral country. Secondly, the right moment for this possibility is now. It is sure that, as soon as shipping will be open again, there will be no possibility whatsoever to get a visa. Thirdly, there is already news that the way from Germany to Lisbon is open. This may possibly not yet be true, but getting a visa takes a while and in the meantime, traveling might become possible again.

Now Placzek, I ask you to help Alex by providing two hundred dollars. This is very urgent. I have written to Viki in great detail and ask you to contact him and not to answer hastily. There are many places where you can borrow the money if you do not have it right now. We have done so as well. The matter is really very urgent. This possibility to help Alex is not in any way risky or compromising to you, so nothing should stop us from saving Alex. Please understand that it depends on you whether he will go to the dogs or be able to live like a human being again, because without your help I cannot help him.

He has asked me to contact you from the beginning, in the first telegram and then over and over again. In the last letter, he writes: "Please give my regards to all the friends, especially to Placzek, who I often think about with affection despite the havoc he has wrought, as well as Viki... I want them all to write me, after so much time this is an extraordinary joy..."

[handwritten additions:] Please send the money to Misi. Once the money has been collected, Alex will be saved. Please think about what you could do, without your help, he will not be able to regain freedom!

Please, hurry up!

Eva

P.S. I have told Viki that you can appoint Gilard or Schnax as trustees to ensure Misi and I use the money correctly if you so choose.

Viki has a copy of this letter.

☙

George Placzek to Eva Zeisel

Ithaca

July 21, 1940

Dear Eva:

Thank you very much for your letter. I have contacted Weisskopf, who is in the Mid-West now, via telegraph. There is another complication to the situation: I have just learned (indirectly, via Switzerland) that my family is in imminent danger and need visas immediately. I'm still wondering how to do this. Nevertheless, I am still willing to go to the utmost limits of the given possibilities, narrowed by these circumstances, for Alex. The day after tomorrow (Wednesday), I am going to New York where I will contact your brother immediately.

Otherwise, I think it will be of little use to our cause if we start to heap mutual recriminations considering Alex's fate over each other. I can very well understand that, behind Russian jail walls, you gradually start to take the abominably wretched parody of a legal process that they play there seriously in every detail and to see the characters of the torturers' tragicomedy as actual reasons or causes for the charge. But, really, by now you know as well as I do that this is not the reality.

Kind regards,
Placzek

☙

Felix Bloch to Eva Zeisel [82]

Stanford, CA

June 9, 1938

Dear Eva,

Congratulations on your timely flight from the Nazi "justice"! My mother had already written to me about it, and I only wished that all my Viennese relatives had also been so sensible, rather than now being in prison or being subject to the evilest extortions. Since the "Anschluss" my correspondence has increased to an extent that is the more frightening when I realize the disparity between my weak strength and the flood of suffering. It was also painful to realize that I could not do anything for Weissberg. After arriving here, I talked to Lauritsen [83] about it, and he explained that the American Embassy in Moscow immediately withdrew from the matter as soon as they learned that Weissberg was a Party member. I don't know whether I could have predicted this, had you told me about it when we met in Vienna. The matter was immediately rubber-stamped as a matter internal to the USSR – from what I know of the present conditions, it would be entirely pointless for Millikan [84] to pursue the matter, and equally pointless if I were to try and persuade him to do that. I can only hope that in Russia that residue of humanity has survived, in which you believe so unshakably. However hard it is for me to acknowledge my powerlessness, I nevertheless believe I have to admit it in verity!

---

[82] Courtesy of Jean Richards and Judith Szapor. Translated from German by Wladimir von Schlippe. (The original letter is available in Bloch's handwriting. Above the address three lines are added:

> I have just received this letter. Alas, there is not much one can do with Lauritsen. All the same, I ask you to pass on to him Einstein's letter and the new one (you will get them from me). I am sorry that the American Embassy has withdrawn!

They are written by a different but unidentified person: whereas Bloch uses the formal German "Sie" to address the recipient, the other writer uses the "Du," reserved to address family and close friends. -WvS)

[83] Charles Christian Lauritsen (1892-1968) was an American physicist at Caltech.

[84] Robert Andrews Millikan (1868-1953) was an outstanding American experimental physicist. Awarded 1923 Nobel Prize for Physics.

I would be very happy to learn that in England you are entering a time of calm and care after the ordeals you had to endure.

With best regards,
Felix Bloch

References

[1] V. Weisskopf: *Mein Leben*, Scherz Verlag, Bern, 1991.

[2] V. F. Weisskopf: *George Placzek, Theoretical Physicist*, Science, 123 (1956), March 9, p. 409.

[3] R. Moore: *Niels Bohr: The man, his science and the world they changed*, (Alfred A. Knopf, New York, 1966).

[4] J. Fischer: *Georg Placzek (1905-1955)*, Čs. čas fyz. A 35 (1985), pp. 607-611.

[5] O. R. Frisch: *What Little I Remember*, (Cambridge Univ. Press, Cambridge, 1979).

[6] D. Cassidy: *Placzek, George*, In: *Dictionary of Scientific Biography 18*, Ed. F. L. Holmes, (Charles Scribners Sons, 1990) pp. 714-715.

[7] J. A. Wheeler (with K. Ford): *Geons, Black Holes, and Quantum Foam. A Life in Physics*, W. W. Norton & Comp.,
http://partners.nytimes.com/books/first/w/wheeler-geons.html

[8] J. R. Goodstein: *A Conversation with Franco Rasetti*, Physics in Perspective 3 (2001), pp. 271-313.

[9] K. Schlüpmann: *Vergangenheit im Blickfeld eines Physikers – Hans Kopfermann 1895-1963*,
http://www.aleph99.org/otusci-index.htm (on Placzek esp. in Ch. II. "Ein 'Aussenminister'," and in Ch. III. "Endstation Heidelberg").

[10] Wheeler J. A.: *Mechanism of Fission*, Physics Today 20 (1967), 11, pp. 49-52.

[11] Williams: *The Development of Nuclear Reactor Theory in the Montreal Laboratory of the National Research Council of Canada (Division of Atomic Energy) 1943-1946.*, Progress in Nuclear Energy 36 (2000), 3, pp. 239-322.

[12] P. R. Wallace: *Atomic Energy in Canada: Personal Recollections of the Wartime Years*, Physics in Canada, March/April 2000, pp. 123-131.

[13] M. Rechcigl Jr.: *Notable American Scientists with Czech or Slovak Roots*, http://members.aol.com/rechcigl/myhomepage/news.html

[14] *Pamětní kniha obce Alexovice, Kroniká*, T. Buček (1923-31), F. Dočekal (1932-48). Archiv Brno-venkov, Rajhrad.

[15] J. Fischer: *George Placzek – an Unsung Hero of Physics*, CERN Courier, 45 (2005), 7, 2005

[16] Personal recollections and archives of local contemporaries, especially, J. Kocourková and M. Nešpůrková.

[17] Čs. čas. fyz. A 19 (1969), 4, p. 515 (interview with Gilberto Bernardini).

[18] G. Placzek: *Rayleigh-Streuung und Raman-Effekt*, Handbuch der Radiologie, 6 (1934), Pt. 2, 205-374.

[19] O.R. Frisch: *The Discovery of Fission – How It All Began*, Physics Today 20 (1967), 11, pp. 43-48.

[20] *Physics in a Mad World*, Ed. M. Shifman, (World Scientific, 2016).

[21] *Collected Papers of L. D. Landau*, Ed. D. Ter Haar, (Pergamon Press, Oxford, 1965).

[22] I. Unna: *The Genesis of Physics at the Hebrew University of Jerusalem.*, Physics in Perspective 2 (2000), pp. 336-380.

[23] Alexander Weissberg-Cybulski, *Hexensabbat*, (Frankfurt, Frankfurter Hefte, 1951, in German), English translation: Alexander Weissberg, *The Accused*, (New York, Simon and Schuster, 1951) and *Conspiracy of Silence*, (Hamish Hamilton, London, 1952). A modern German edition was released relatively recently in Austria under the title *Im Verhor: Ein Überlebender der Stalinistischen Sauberungen Berichtet*, (Europa Verlag, 1993).

[24] G. Gorelik, V. Ya. Frenkel: *Matvei Petrovich Bronstein and Soviet Theoretical Physics in the Thirties*, (Birkhäuser, Basel, 1994).

[25] H. Bethe, G. Placzek: *Resonance Effects in Nuclear Processes*, Phys. Rev. 51 (1937) 450.

[26] K. Bird, M. J. Sherwin: *American Prometheus: The Triumph and Tragedy of J. Robert Oppenheimer*, (Knopf, 2005).

[27] P. Goodchild: *J. Robert Oppenheimer – Shatterer of Worlds*, (Fromm International Publishing Corporation, New York, 1985).

[28] J. Robert Oppenheimer to K. D. Nichols, March 4, 1954. Nuclearfiles.org – a project of the nuclear age peace foundation.
http://nuclearfiles.org/redocuments/1954/540304-opp-nicholsprint.htm

[29] A. Pais: *Niels Bohr's Times, In Physics, Philosophy, and Politics*, (Clarendon Press, Oxford, 1991).

[30] W. Lanouette: *Genius in the Shadows. A Biography of Leo Szilard, The Man Behind the Bomb*, (The University of Chicago Press, Chicago, 1992).

[31] D. C. Fakley: *The British Mission*, Los Alamos Science, Winter/Spring 1983, pp. 186-189.

[32] G. C. Laurence: *Early Years of Nuclear Energy Research in Canada*, Atomic Energy of Canada Limited, May 1980.
http://www.cns-snc.ca/history/early_years/earlyyears.html

[33] A. Weinberg: *The Second Fifty Years of Nuclear Fission*,
http://www.cns-snc.ca/history/fifty_years/weinberg.html

[34] H. Brooks: *Autonomous Science and Socially Responsible Science: A Search for Resolution*, Annu. Rev. Energy Environ. 26 (2001), pp. 29-48.

[35] E. Segrè: *A Mind Always in Motion: The Autobiography of Emilio Segrè*, (University of California Press, Berkeley, 1993)
http://ark.cdlib.org/ark:/13030/ft700007rb/

[36] IAS – A Community of Scholars,
http://www.ias.edu/people/cos/users/gplaczek01

[37] E. Amaldi: *George Placzek*, La Ricerca Scientifica 26 (1956), pp. 2037-2042.

[38] Victor Weisskopf, *The Joy of Insight*, (BasicBooks, 1991), pp. 68-69.

[39] I. V. Obreimov, *Fifty Years of Development of Natural Sciences*, Report of first UPTI

Director at the Meeting of the Scientific Council devoted to UPTI's 40-th Anniversary, Vestnik Academii Nauk USSR, 1971, No. 10, in Russian.

[40] L. Landau and G. Placzek, *Struktur der Unverschobenen Struelinie*, Physikalische Zeitschrift der Sowjetunion, **5**, 172 (1934). Received by the Editorial Office on November 14, 1933.

[41] Eva Zeisel, *A Soviet Prison Memoir*, Ebook, (evamemoir.com) (Amazon Kindle Edition, 2012).

[42] L. Landau and E. Lifshitz, *Electrodynamics of Continuous Media*, (Butterworth-Heinemann; 2nd Edition, 1984) pp. 431-432.

[43] M. Shifman, *Standing Together in Troubled Times*, (World Scientific, Singapore, 2017).

[44] *Four Interviews* , Eds. Yuri Ranyuk and Yuri Freiman (NNTs KhFTI, Kharkov, 2010).

[45] Edward Teller (with Judith Shoolery), *Memoirs*, (Perseus Publishing, Cambridge, 2001), pp. 78 and 181.

[46] L. A. Kochankov, *Atom Intelligence and KB-11*, (Russian Federal Nuclear Center VNIIEF, 2011), in Russian.

[47] Felix Bloch to Eva Zeisel, Stanford, June 9, 1938, Eva Zeisel Papers.

[48] Aleš Gottvald, "Kdo byl Georg Placzek (1905-1955)," *Čs. čas. fyz.*, Vol. 55, No. 3, 2005, pp. 275-287.

PART 2

NKVD FILES OF KONRAD WEISSELBERG

Chapter 2

NKVD Files of Konrad Weisselberg

As was mentioned in Chapter 1, previously unavailable information about Georg Placzek's second visit to Ukrainian Physical Technical Institute in Kharkov in 1936-37 came from an unexpected side: the arrest and interrogation documents of Konrad Weisselberg. They were kindly provided to us by his grandson Art Kharlamov. This file *per se* is a remarkable human document that teaches us – better than any textbook – that horrible atrocities are committed in the name of allegedly noble ideology. And they are not an issue of the past; they continue to be committed today.

Weisselberg's file has never been published in English, therefore we decided to present it here in its entirety rather that limit ourselves to a number of quotations.

A few words about Konrad Weisselberg.[1] Born in 1905, Konrad Weisselberg graduated from Vienna University and then earned there his doctorate in chemistry. By the early 1930s, his elder sister Erna and Konrad himself were members of the Communist Party. His elder brother Marcel joined their father's business.

Konrad Weisselberg arrived in Kharkov in 1934 at the invitation of Alexander Weissberg, who found him a research position at the Coal and Chemistry Institute. In 1936 he was hired by UPTI and

Figure 2.1   Konrad Weisselberg.

became Weissberg's neighbor. In Weissberg's *The Accused* [1] Konrad

---

[1]See Alexander Weissberg's book [1] and the Collection [2]. Konrad Weisselberg's photo on the right was presumably made in Kharkov. Courtesy of Mica Nava and Katerina Sajez.

Weisselberg appears as Marcel.

He married a Ukrainian girl, Anna Mykalo (also known as Gala or Galia) (Lena in Weissberg's *The Accused*), but this did not help him. In February 1937, Konrad Weisselberg renounced his Austrian citizenship; he was arrested on March 4, 1937. Weisselberg was indicted on charges of sabotage against the Soviet Union, establishing a criminal contact with the German intelligence agent A. Weissberg, and becoming a member of Weissberg's counter-revolutionary group. After ten months of interrogations, the Procurator General of the USSR, Vyshinsky, sentenced him to death without trial on October 28, 1937; he was executed on December 16 of that same year [3].

Igor Starostin recollects [2]:

> My mother told me that after Konrad Weisselberg's arrest, aunt Galia [Weisselberg's wife -MS] was allowed to bring some food for Konrad to the NKVD; in return she was given Konrad's underwear for washing. The clothes were covered with blood. Then one day, a prison official did not accept food and clean clothes from aunt Galia; she was told that it was no longer necessary.

THE FILE OF KONRAD WEISSELBERG [2]

Transcribed and commented by his son Alexander Kharlamov from files held in the Kharkov KGB archive.

A SON REGAINS HIS FATHER

Alex Kharlamov

I'm writing to you in Russian, as we agreed when we spoke on the phone. A new publishing venture by the name of 'Rehabilitated by History' has recently been started in Kharkov, and they intend to publish a multi-volume edition dedicated to victims of the years of communist repression. They can obtain files from KGB archives. On December 16, 1996, I managed to see and make a copy of my father's file. It was 59 years to the day since he was shot. At the very moment I picked up the file a violent snow-storm erupted, accompanied by thunder. A thunderstorm in winter? Most unusual in our part of the world! For me, Kiffer, it was a most eerie sensation. Reading my father's file has been a harrowing experience. At first my father's file was part of the Weissberg case; at a later stage the two cases were separated, probably because it was realized that Konrad had taken Soviet citizenship, whereas Weissberg had not. What follows is an exact transcription of my father's data with all its mistakes, inaccuracies and strange turns of phrase left unchanged.

---

[2] Konrad Weisselberg (1905-1937) was arrested by the NKVD in Kharkov in 1937. The records of his interrogations below contain valuable information about Placzek's second visit to Kharkov (which is inaccessible otherwise). The first publication of the Russian original by Yuri Raniuk can be found in the book Yu. V. Pavlenko, Yu. N. Raniuk, and Yu. A. Khramov, *The Case of UPTI*, (Feniks, Kiev, 1998, 324 pp.), Chapter 4. The English translation (with commentaries) due to the Kharlamov family is published here with kind permission of Art Kharlamov, Konrad Weisselberg's grandson. He told me that he remembers some of the "Placzek stories" as they were narrated to him by his grandmother, Anna (Galina) Mykalo, see p. 121. -MS

ARCHIVE DOCUMENTS

CRIMINAL CASE

Weisselberg K. B.
Kharkov District Ukrainian KGB Archive No 016138
Ukrainian Council of Ministers (USSR)

DECLARATION

Kharkov. March 4, 1937

I, Head of section 2 of the third division, first lieutenant Tyshkovsky, having studied the investigatory material accusing citizen Weisselberg, Konrad Bernardovich (born 1905, Barit [sic], Rumania, Austrian by birth but now a Soviet citizen, until recently employed as a scientist at the Coal and Chemistry Institute, Kharkov, not a Party member) of crimes as described in Articles 54-41, 54-11 of the Penal Code, to wit that he arrived in the Soviet Union in order to conduct sabotage against the Soviet Union and that he worked with others to this end: have concluded:

that Weisselberg arrived in the Soviet Union in 1934 in order to conduct sabotage against the Soviet Union. In his criminal activities he worked with others: have decided: that Weisselberg K. B. be arraigned under Articles 54-41, 54-11, to inform the public prosecutor accordingly and to keep him under arrest. Signed: Tyshkovsky, Arrov, Shumskii.

This declaration has been shown to me.
Signed Weisselberg, March 11, 1937.

March 4, 1937: I, Vazonov, deputy public prosecutor for the Kharkov district, having studied the material prepared by the third division of Ukrainian KGB, Kharkov district, on the counterrevolutionary fascist and sabotage activity of Weisselberg K. B., scientist at the Kharkov Coal and Chemistry Institute, find:

that the investigators have gathered sufficient evidence to prove Weisselberg guilty of anti-state activities: I agree to their application and sanction the arrest of Weisselberg K. B.

Ukrainian SSR
Peoples' Commissariat of Internal Affairs

Kharkov District Administration
Order No. 231, issued March 4, 1937, valid for 48 hours.

Comrade Drescher, 2nd lieutenant of the Kharkov district NKVD, has been instructed to conduct a search and arrest of citizen Weisselberg K. B., of Kharkov, Tchaikovsky St, House 16, Flat 8. All organs and citizens of the Ukrainian SSR are obliged to assist the officer bearing this order in the execution of his duty.

Signed by the head of the Ukrainian NKVD, Kharkov District, Captain (illegible), and department chief, lieutenant (illegible)

Record of search proceedings:

Present during the search: Mykalo Galina (wife).

Details of arrestee:

1. Name: Weisselberg

2. Konrad Bernardovich

3. dob October 25 , 1905

4. Place of birth: Berlad, Romania

5. Address: 16/8 Tchaikovsky Street, Kharkov

6. Profession: chemist

7. Place of work: Coal and Chemistry Institute, Kharkov.

8. Passport issued: -

9. Social background: father is a timber merchant with his own business in Vienna.

10. Social status: pre-revolution, student; worked to age 30.

11. Education: higher

12. Party status: non-party member; was a member of the Austrian Social Democratic Party from November to July 1934, member of the German Communist Party.

13. Nationality: Soviet citizen, Austrian Jew.

14. Liability for service in the Armed Forces: -

15. Service in the White Army and others: -

16. Family: wife Anna Davidovna Mykalo, aged 27; son, Alexander, March 4, 1937

Extract from the declaration in the Ukrainian language, witnessing the granting of Ukrainian citizenship to Weisselberg K. B., signed

Zinchetko, secretary; March 3, 1937.[3]

RECORD OF INTERROGATION – Weisselberg K. B., March 9, 1937

Q. Tell us about your background.

A. I was born in 1905 in Berlat, Rumania, to a timber merchant's family. I have an elder brother and two sisters. My brother worked with my father in the business and is now a timber merchant in his own right. In 1907 we left Berlat for Chernovtsi, Bukovina, now Austria. In Chernovtsi I attended elementary school until 1914, when Russian forces invaded the town and we moved to Vienna, where I finished my primary and then secondary education in 1925.

In 1922 I joined the "Pathfinder" sports club, which was completely unpolitical. It was purely for sports. In summer the club-members went to a summer camp in the country. At one of these camps I met a young Czech by the name of Danny Gerdo. Gerdo was a communist agitator. When I got to know him better he gave me a book called *The Communist Alphabet* which I read and returned.

In 1922 I joined the Union of Underground Communists in Vienna. At the time the leader of the Union was Genya Lange (Genya Kvinter) who's now living in Moscow. I took an active part in the organization until 1925.

I had joined the School-Students' Union with a friend of mine, Comrade Gustav Deutsch, who is now a member of the Austrian Comintern. He's the son of an important Austrian Socialist politician, now a general in the Spanish Army. Gustav Deutsch now lives in Eltsy [a town in the Soviet Union]. In 1925 I transferred to the Socialist Students' Union, which was under the influence of the Social Democratic Party. I was a member until 1930 and head of the Socialist Chemists' Union. I was arrested twice after street-fights with the fascists. In 1928 I joined the Austrian Social Democratic Workers' Party and remained a member until 1934. I graduated from university in 1934 and got a job as a chemical engineer at an oil refinery in Drening, 60 km from Vienna. I joined the SDP.

In 1933 I joined the Austrian Communist Party. My political views were very much influenced by my sister Erna. She was a member of

---

[3]Two days before the arrest.

the Austrian Comintern and often gave me communist literature which I distributed to workers at my factory. In 1934 there was a fascist putsch in Austria. I was in Vienna then but I didn't take any active part in opposing the fascists. I carried on working in Drening. I was busy setting up communist cells in the factory and also kept up my connections with the communist center in Vienna.

On the first of May 1934 I was searched on suspicion of possessing illegal weapons and communist literature. I was fined on the weapons charge. In December 1934 I was searched again, on suspicion of possessing communist leaflets. And in fact I had, together with my cells, distributed a large number of Comintern leaflets on May Day. Still in 1934, I was ordered by the Comintern to help two people unknown to me make illegal border-crossings. My job was to take them to the Czech border on my motorbike.

In 1933 I visited the USSR as a tourist and in Kharkov I met up with an old friend of mine from the Socialist Students' Union, Alexander Semeonovich Weissberg. I spent a fortnight in Kharkov and Moscow and returned to Vienna, having arranged with Weissberg and my present wife, Mykalova Galina Davidovna, that they would invite me back to the USSR. Weissberg worked at the Ukrainian Physical and Technological Institute (UPTI) and my wife-to-be was his housekeeper.

In July 1934 I was invited to come to work at the Coal and Chemistry Institute (CCI) in Kharkov. In August 1936 I moved to UPTI. In January 1935 I went on leave to Vienna. In January 1936 I decided to take Soviet citizenship and went back to Vienna for a fortnight to settle my affairs there. In January 1937 I was granted Soviet citizenship.

From 1926 to 1930 I was a member of the Schutzbund in Vienna.

Q. In 1934, after you were searched in Drening, were you prosecuted for possession of illegal weapons and literature? Were you punished?

A. No, they found nothing incriminating.

Q. This was after the fascist putsch in Austria?

A. Yes, shortly after they seized power.

Q. In other words, the fascists didn't prosecute you for the possession and distribution of communist literature?

A. No, they hadn't started prosecuting people for that – yet.

Q. Did they find anything the second time you were searched, in May 1934?

A. No, nothing.

Q. Which of your Austrian friends are living in the USSR now?

A. The following:

1) Goldschmidt, Hans – a member of the Austrian Communist Party. He lives in Moscow and works for the publishing firm "Foreign Worker." We've known each other since SSU days.

2) Berdl Graf – his real name is Brodo. Member of the Austrian Communist Party, lives in Moscow, we met in the SSU.

3) Edel. A doctor, member of the ACP, a friend of Goldschmidt's, he's known to me since the SSU.

4) Tumheim – member of the Schutzbund. He lives in Kharkov in the Hotel *Spartak*. Another friend from SSU days.

5) Glesner, Martin. A senior lecturer at the Academy of Science, not a party member, lives in Moscow, a friend from Vienna.

6) Golmend, Hans. An agronomist, not a party member, he lives in a village called Bolshoi Kolodets here in the Ukraine. I met him at the SSU in Vienna.

7) Golmend, Otto. He's an engineer, lives in Kramatorsk, I've known him since Vienna.

8) Ernst Wieden. Head of the Austrian section of the Comintern. Lives in Moscow.

9) Schindler, Julius. He worked with me at the factory in Drening. He was a member of the Schutzbund and lives in Moscow. He's a member of the Hungarian Communist Party.

10) Rubinstein, Gregory. He's an economist, lives in Moscow and I've known him since Vienna.

11) Strieker, Michael. Works in film production in Moscow, lived in Vienna.

I want to add that my wife's brother, Grigori Mykalo, lives in Moscow and works in the NKVD there, and I know him very well.

Record of Interrogation by the Head of Section 2 of the Third Division, Tyshkovsky. March 16, 1937

Q. When and why did you come to the Soviet Union?

A. My first visit to the Soviet Union was in May 1933, as a tourist. I wanted to find out if I could get a job here. I spent 14 or 16 days in Kharkov and Moscow. My second visit was in July 1934, when I was invited to work as a scientist at CCI.

Q. You went to work at CCI on a contract?

A. No, I was invited by the Director, Meier.

Q. I'd like some more details about your arrival in the USSR.

A. From 1932 on I was in touch with my friend Weissberg, who was living in Kharkov. I asked him to try to find me work in the Soviet Union. I discussed this with Weissberg when I came here in May 1933. Weissberg talked to Zuckerman, deputy director of CCI, Straller, head of the laboratories there, Meier and others. Before I went back to Austria in 1933 Weissberg promised to find me a job in the Soviet Union. I stayed in touch with Weissberg and in July 1934, quite unexpectedly, received the invitation from Meier to come to Kharkov.

Q. Did Weissberg get in touch with anybody in Moscow about your coming to the Soviet Union?

A. He did get in touch with the Ministry of Heavy Industry, but who in particular he talked to, I don't know.

Q. You lived the whole time in Weissberg's flat?

A. Yes, from the moment I arrived here I lived in Weissberg's flat. In his invitation Meier explained that I wouldn't be allocated a flat in Kharkov.

Q. What foreigners came to see you in Kharkov?

A. In 1936 Placzek, a Czech, came to visit Weissberg. He stayed with us for about 6 weeks. There was also Victor Weisskopf, an Austrian who'd come from Copenhagen. He and his wife stayed with us for 5 days or so. In 1935 two Austrians arrived from Czechoslovakia, one was called Moser, the other one's name I can't remember. They stayed with us for about 10 days.

Q. Who was this Placzek?

A. Placzek is a well-known physicist who studied in Vienna. He's visited many countries in Europe and America. I got to know him in Kharkov in '33 during my first visit to the Soviet Union, at Weissberg's flat. Up till 1934, when he came to the Soviet Union, Placzek lived in Copenhagen where he was an assistant to Professor Bohr. He came to keep in touch with Soviet physicists.

Q. What was the link between Weissberg and Placzek?

A. They had a good relationship. They'd known each other for a long time. I don't know what link there was.

Q. Do you know Placzek's political views?

A. I don't think he belonged to any party. He certainly never talked about such things. Once he told me that the Soviet government had fixed the trial of the Trotskyists and that German aggression is aimed not at the Soviet Union but at France. Interestingly enough Placzek said that he'd had a meeting with Bukharin[4] in Moscow. I don't know what went on at that meeting, but I had the impression that they'd got on well. It took place when Bukharin was still editor of Izvestia. I should add that I don't know Placzek well and I can't judge as to any anti-state activities on his part.

Q. Did Placzek discuss the Trotskyists' trial with you alone?

A. I'm not sure, but I think Ruhemann and Houtermans told me that he told them, too, that the trial was fixed.

Q. Did Weissberg know about Placzek's political leanings?

A. If I knew about them, then he must have known too.

Q. What did Placzek tell you about Trotskyists abroad?

A. We never discussed the subject.

Q. Did you write to Placzek after he left?

A. When Placzek left Kharkov for Moscow I sent him a letter of a private nature, asking him to write from Copenhagen to my father in Vienna, to tell him that I'd been out of work for 6 months and that I wanted to come back to Vienna with my wife to find a job there.

Q. Why did you have to get in touch with your father in this way? Couldn't you have written directly from Kharkov?

---

[4]See footnote on page xv.

A. I knew that the Austrians checked all letters coming from the Soviet Union. I didn't want the Austrian authorities to know about my situation or about my intention to return to Vienna.

Q. You're not giving us a straight answer. Explain why you had to use such a means of communication with your father.

A. I stand by what I've said – I had no other motives.

Q. When did you write to Placzek in Moscow?

A. In early January 1937.

Q. Why did you lose your previous job?

A. On August 15,1936, I resigned from CCI – but this was a mere formality. In fact there were staff cuts. Later I worked at UPTI as a consultant, but without a permanent job. That came to an end in December and I had no other prospects, so I wrote to Placzek and soon afterwards got a temporary job in the Coal And Chemical Construction Bureau.

Q. Did you get a reply from your father?

A. No, I didn't. Placzek wrote from Moscow that he would do as I asked and write to my father from Copenhagen.

Q. Did you always write to your father in this way?

A. No, that was the first time.

Q. How many times did you go back to Austria after your second visit to the Soviet Union?

A. Twice. The first time was in August 1935, when I took leave due to me in Austria-Hungary; the second was in January '36. I went to Austria to settle my private affairs when I applied for Soviet citizenship.

Q. Who did you meet on those two visits?

A. Only friends, I didn't meet any officials.

Q. Where did you get your passport for foreign travel?

A. In summer 1935 the Austrian authorities decreed that anyone wanting to travel abroad must register their passport with the police; so after my holiday I went to the police and my passport was extended to 1940. On my second visit, in 1936, I received a Viennese police visa for travel to the Soviet Union.

Q. When and where did you get your first passport?

A. I got my first one in 1925 when I went to Italy. I was going to use the same one for a trip to Athens but it was stolen in Trieste, by a White Russian emigré who was later arrested. I don't know his name, but he was arrested in Vienna for theft. I got my last passport in about 1930.

Q. Have you visited the Austrian Embassy in Moscow?

A. No, never, because my passport was valid until 1940.

Q. Have you visited the German consular department in Kharkov?

A. No, never.

Q. Do you have any acquaintances among foreign diplomats in the Soviet Union?

A. No.

I have read this record and confirm that it is accurate.
Signed (Weisselberg)
Interrogator: Tyshkovsky

Record of interrogation of Weisselberg, March 20, 1937

Q. Why did you leave Austria when you had a good job and you were working for the party?

A. As a Jew I had no prospects, and besides, I was frightened of being arrested.

Q. Why did Placzek come to the Soviet Union?

A. To keep up his contacts with Soviet physicists – Landau, Frenkel, Tamm, Beck.

Q. Where did Placzek live in Kharkov?

A. He lived with us in Weissberg's flat, although UPTI had got him a place in a hostel.

Q. Why did Weisskopf come to the Soviet Union?

A. For the same reason as Placzek. He stayed with us. He was a good friend from the old days of the Socialist Students' Union - that's where I met him and Weissberg. Moser stayed in Weissberg's flat too.

Q. What do you know of the connection between Weissberg and Rataichik?

A. That Weissberg stayed with him when he was on OSGO business in Moscow.

Record of interrogation of Weisselberg, March 22, 1937

Q. You worked in Austria until your departure for the USSR?

A. I worked at Drening, near Vienna, at the Kreditel Mainer oil-refinery.

Q. What did you earn there?

A. I got 250 schillings a month. I had no financial worries because my father was a wealthy man and helped me.

Q. In your earlier answers you said you were a member of the Austrian SDP and in 1933 you joined the Austrian Communist Party. When exactly did you join?

A. I joined the Austrian CP in December 1933. Before that I was a member of the SDP.

Q. What was your job in the Party?

A. I organized a communist cell in my factory and was the local representative of the Party. I kept in touch with the central organs of the Party and obtained propaganda material.

Q. When exactly did you arrive in the Soviet Union?

A. On July 18, 1934.

Q. Did you get permission from the Austrian CP for your departure?

A. I didn't have time to inform them. I left without permission.

Q. In other words you came to the Soviet Union although you had a job in Austria, and abandoned your Party job without permission? Give us your real motives for coming here.

A. I wanted to go to the Soviet Union because I am a Jewish Socialist and I saw no prospects for a career in science in Austria. In addition my flat was searched twice in 1934, so the police were aware of my communist activities. I was sure I'd be arrested within two or three months. That's why I came to the Soviet Union, in the hope that I could get retrospective permission for my departure. I wasn't aware of party discipline in this matter. I assumed that because I'd left the country I'd left the Party too.

Q. What was the purpose of Placzek's visits to the Soviet Union?

A. As I said, Placzek visited the Soviet Union twice, in 1933 and '36, in both cases to visit the famous Soviet physicists, Frenkel, Tamm and Beck.

Q. Where did Placzek stay during his visit to Kharkov?

A. He stayed with us at Weissberg's flat although UPTI gave him room in their hostel.

Q. How long has Weissberg known Placzek?

A. I'm not sure when they met but it was abroad, well before Placzek's visit in 1933.

Q. Why did Weisskopf come to the Soviet Union?

A. For the same reason as Placzek. He stayed at Weissberg's flat. Weissberg is a good friend from our university years. We studied together.

Q. Why did Moser and his friend visit the Soviet Union?

A. Moser and his friend, whose name I can't remember, stayed at Weissberg's. I don't know the reason for their visit. Moser is a former Social Democrat; he emigrated to Czechoslovakia. Weissberg told me that they came to the Soviet Union to negotiate with the Comintern about joining the Communist Party here.

Q. What do you know about chief engineer Zuckerman of OSGO and of his connections with Weissberg?

A. About Zuckerman. I only know that before he came to work in Kharkov he worked in Lisichansk. I don't know when he got his job with OSGO. Zuckerman and Weissberg had a purely professional relationship. Zuckerman would come to Weissberg's flat and talk only about professional matters.

Q. What do you know of the relationship between Zuckerman and Pushin?

A. I have no idea. This is the first time I've heard the name Pushin.

Weissberg told me that he visited Pushin once or twice on business in Moscow and Weissberg thought he was a very nice chap. Weissberg couldn't imagine that Pushin might be involved in the kind of thing we read about in the newspapers.

Q. What do you know of the relationship between Weissberg and Rataichik?

A. Weissberg told me he once visited Rataichik on OSGO business. The conversation had something to do with the trial. Weissberg said he didn't like Rataichik.

Answers recorded correctly. (Signature).

Record of Interrogation April 5, 1937

Q. Tell us about the anti-Soviet conversations that Placzek initiated.

A. Placzek mainly claimed that the trials of Zinoviev and Kamenev were fixed – show trials, in effect – and that the use of drugs against the accused by the authorities proved the point. When Placzek talked in this way he would become very agitated and passionate, not only with me but with Weissberg and lots of other people. He would discuss this with Ruhemann, Houtermans, their wives, and with one of the UPTI communists as well. Once Weisskopf came when Weissberg was away and rebuked Placzek for talking in this way – he was frightened of exposure and the inevitable unpleasant consequences. Once Placzek said that German aggression was aimed not at the Soviet Union but at France. All this was said in the presence of Weissberg.

Q. Did Placzek stay at your flat on his last visit in December 1936?

A. After Weissberg returned from Moscow, Placzek moved out of our flat and into the hostel but came back a day or two later.

Q. Was there some other reason for Placzek's move to the hostel?

A. No, I'm sure of that. Placzek didn't want to be a burden because of the lack of space, but Weissberg didn't mind him being there.

Q. Did Weissberg ever tell Placzek to leave the flat because of his anti-Soviet opinions?

A. No, Weissberg simply warned him about this. He told him to stop talking in this way or else he would have to ask him to leave our flat. But Placzek only moved out temporarily.

Q. When did Placzek tell you about his visit to Bukharin?

A. A few days after he arrived in Kharkov in December 1936. Placzek told me that he had stopped in Moscow on the way and visited Bukharin. He didn't tell me the reason for the visit but I got the impression that Placzek had known Bukharin well before he came to the Soviet Union. Apparently they talked about Italian fascism. I do remember there was a third person present during my conversation with Placzek, but I can't remember whether it was Weissberg or someone else.

Q. Who did Placzek meet in Kharkov?

A. His main contacts were with some of the UPTI scientists, with Professor Landau in particular, whom he met at our flat as well as at the Institute, a group of Landau's graduate students, and Rozenkevich. As far as I know he had several meetings with Rozenkevich. Placzek met the whole group of foreign specialists at the Institute.

Record of Interrogation of RUDOLF ANDERS, April 5, 1937

Rudolf Anders, otherwise known as Erwin Kohn, is deputy head in charge of construction at Zaporozh-Stal [a huge steel-works]. He gave evidence against Weissberg, Weisselberg and Placzek. He visited Kharkov occasionally.

In 1932 Weissberg visited a number of *kolkhoz* [collective farms], mainly German, in the Dnepropetrovsk district, to find out for himself the true situation there. He did this of his own accord. In the summer of '33 he shared his impressions of his trip, which amounted to his view that these villages were in the grip of famine. There was a critical lack of manpower, and many people, not just kulaks, were being persecuted. The reason for all this, according to Weissberg, was that collectivization was being imposed too early and that the expropriation of the kulaks had gone too far. This conversation took place in Kharkov, in Weissberg's flat, in 1934. Weissberg rarely criticized credit and monetary policy in the planning of the economy; he supported the line of the right-wing opposition on this matter, which can be summed up as the idea that, at this stage of the development of the Soviet Union, state capitalism was the only possible way forward. When it came to discussing international politics, Weissberg took a rather defeatist attitude. He insisted that once the fascists came to power in

Germany, fascism would spread throughout Europe. He predicted that by 1938 a new anti-Soviet fascist bloc would have formed which must result in the destruction of the Soviet Union. This is a good illustration of Weissberg's right-wing counter-revolutionary opinions in recent years. His right-wing attitude is also confirmed by his connections with Karl Kraab, the leading rightist expelled from the German Communist Party, and with a number of Austrian Social Democrats. Because of his senior position in UPTI Weissberg must have had meetings with Bukharin. I emphasize that in 1934/35 Weissberg was head of construction at OSGO. This [hydro-electric[5]] station is a very important defense priority because of its connection with the nitrogen industry. There's one episode in particular I want to mention. Sometime between December 15-20, 1936, Weissberg came to my flat in Moscow and told me his wife Striker had been arrested by the NKVD in Leningrad and that he thought all foreigners should leave the Soviet Union immediately. Trying to convince me of this he said that he would be leaving as soon as possible. As another example of the harassment of foreigners he mentioned Konrad Weisselberg, who had lost his job at CCI (according to Weissberg) simply because he was a foreigner and now couldn't find any work. I learnt another interesting detail about Weissberg from Lotte Schwartz. I know Weissberg as a specialist in engine-construction, particularly, I think, aviation engines, whereas since 1930 Weissberg has been at UPTI working as a physicist.

Record of Interrogation May 11-12, 1937

Q. Do you know Gustav Deutsch?[6]

A. Yes, I've known him since 1921, when we were members of the school-students' union and later the SSI and the Schutzbund. I was a member of the Schutzbund until 1930.

Q. What do you know of Deutsch's activities abroad?

A. Gustav Deutsch is the son of a former war minister in the Reiner government in Austria. In about 1925 Deutsch joined the Austrian Social Democratic Party, where he was a moderate, unlike his father who

---

[5] In fact that was a cryogenic facility.

[6] Gustav Deutsch (1906-1938) was a Viennese engineer who emigrated to the USSR on April 12, 1934, and resided in Voronezh and Elets. He was arrested in Voronezh on October 21, 1937, on the charges of espionage and involvement in a terrorist organization. He was sentenced to death by the Military Collegium of the Supreme Court of the USSR on June 14, 1938, and executed by firing squad on the same day in Moscow.

was on the reformist right. In 1929 Deutsch graduated as a construction engineer from the Technical High School in Vienna. He moved to France where he worked in a small provincial city and later in Paris. In Paris he married a Swiss girl, Agnes – I don't know her maiden name – who worked in an office and occasionally reported for foreign newspapers. In 1932, when foreigners were finding it hard to get work in Paris, he returned to Vienna and his father helped him get a job in his field in the Viennese local government Commune. Deutsch's father was the leader of the Schutzbund uprising in Vienna in February 1934. After its defeat he fled to Czechoslovakia and later travelled to America, England and other countries to lecture about the uprising. At the moment he's in the Spanish Republican Army. Gustav himself didn't take part in the uprising – I don't know why – or at least not in the armed struggle. After the uprising was suppressed (around 20 February 1934) he stayed on in Vienna, but one day he came to see me and said that as the son of Reiner's former War Minister he should probably quit Austria to escape persecution. Shortly afterwards he did indeed leave for Prague and settled in the Soviet Union in April 1934. To begin with he lived in Voronezh where he worked on the Moscow-Donbass project and moved to Yelets in 1936. Agnes works as a journalist, occasionally writes for the "Foreign Worker" publishing house and also reports for the French and Swiss press.

Q. Do you know why Deutsch didn't take part in the Schutzbund uprising?

A. I don't know exactly, we never talked about it, but from our political discussions then and earlier I knew that he was against the armed uprising.

Q. Have you met Gustav and Agnes Deutsch in the Soviet Union?

A. I met them two or three times between 1934 and 1936. The first time they came to see me was in 1934 for the anniversary of the October Revolution. They spent a couple of days with me. In 1934 they stopped over for a day on the way to Crimea for a holiday, and in August 1936, when I was in Moscow trying to sort out my sacking from the CCI, I met them both in the VTSSPS [Soviet Centra Trade Union] hostel. I told them about losing my job and why I was in Moscow. I also remember that in 1936 Agnes spent three days in Kharkov. It was a business trip – "Foreign Worker" had commissioned her to study the situation of the former Schutzbund members in the Soviet Union.

They intended to publish a pamphlet unmasking the vicious lies of the Schutzbund "returnees" from the Soviet Union. She stayed in a hotel and visited me a couple of times. I never heard a single anti-Soviet remark from either Agnes or Gustav.

Interrogation conducted by (signature).

Extract from the statement by accused prisoner Valentin Petrovich Fomin, September 30

... The Gestapo organization within UPTI.

<u>Weissberg</u>: Link to the German consulate. Organizer. Link to the Trotskyists (Pyatakov).

<u>Fomin</u>: Supposed to help Houtermans recruit members among the young. Link with the Kharkov Electro-Mechanical Works. Minor sabotage at UPTI to hamper progress.

<u>Weisselberg</u>: Work at CCI (espionage).

Statement confirmed by Sergeant (signature)

Record of Investigation, June 7, 1937

Q. The investigators have information that you are part of the counter-revolutionary right-wing group formed by Weissberg at UPTI.

A. I have never been part of any counter-revolutionary group and have no idea that any such group existed.

Q. So you claim that you have never been a member of any counter-revolutionary group?

A. Yes, I do.

Q. And you also claim that Weissberg did not form such a group at UPTI?

A. Yes, I'm sure there was no such group.

DECLARATION

The prisoner Weisselberg is being kept in Kharkov prison. There is no material evidence of his guilt and from today his case is transferred to the Military Investigation Board. Signed by Weissband, supervisor, 3rd Department UKGB.

DECLARATION

I approve: signed Reichmann.

The accused has only admitted to taking part in anti-Soviet discussions with the physicist and Trotskyist Placzek. The testimony of Anders, and partly that of Weissberg, exposes him as a member of a counter-revolutionary Trotskyist group. Fomin's testimony reveals him to be a German spy. This Trotskyist spy has not admitted his guilt. Signed: Tomuyev, Weissband.

SUMMARY OF CASE NO. 68020 re WEISSELBERG K. B.

Accused of violations of Articles 54-4, 54-6, 54-11 of the Ukrainian Criminal Code.

The 3rd division of the Kharkov NKVD has evidence that a scientist at UPTI, Weisselberg K. B. is a member of the Trotskyist sabotage group which existed at UPTI; and that he has engaged in espionage. He was therefore arrested and called to account for his actions.

The investigation uncovered the following:

Weisselberg, son of a wealthy Austrian timber merchant, was a member of the bourgeois youth organization "Polatfinder" [sic: should be Pfadfinder] from 1920 to 1925. From 1925 to 1928 he was a member of the Socialist Students' Union. From 1928 to 1934 he was a member of the Austrian Social Democratic Party. He arrived in the Soviet Union in 1934 and took up Soviet citizenship in 1937. On his arrival in the Soviet Union in 1934 Weisselberg contacted Weissberg (under arrest), who is an Austrian citizen. They have known each other since their Austrian days. Weissberg is a proven agent of German intelligence. According to the testimony of the Trotskyist Anders, arrested in Zaporozhiye, Weisselberg is a member of the counter-revolutionary group of specialists at UPTI. The head of the group is the spy Weissberg. In his evidence Weissberg confirmed the existence of the counter-revolutionary group and pointed out that Weisselberg was closely involved with it. Weisselberg himself admitted to taking part in counter-revolutionary discussions with the Trotskyist physicist Placzek, who visited the

Soviet Union and told Weisselberg of his connections with Bukharin. According to the testimony of the German spy Fomin (under arrest) Weisselberg was part of an organization involved with espionage for Germany and was active during his period of employment at the Coal and Chemical Institute. Weisselberg himself only admitted to taking part in anti-Soviet conversations with the physicist and Trotskyist Placzek. Testimony from Anders and partly from Weissberg reveals him as a member of a counter-revolutionary group. The evidence of Fomin exposes him as a German spy.

On the above grounds:

Weisselberg K.B., dob 1905, place of birth Berlat, Rumania, Jewish, son of a wealthy Austrian timber merchant, not a Party member, Soviet citizen, member of a bourgeois youth organisation from 1922 to 1925, SSU member 1925/28, Austrian SDP member 1928/34, employed as a scientist at UPTI before his arrest:
is accused of membership of a Trotskyist group and involvement in espionage, in other words of crimes defined in Articles 54-4, 54-6, 54-11 of the Penal Code of the Ukrainian SSR.

On the above grounds:

It has been decided to send Special File 6805 on Weisselberg K. B. to the administrative division of the UKNKVD in accordance with decree no. 0485 of NKVD, USSR.

Signed: Weissband.

Approved. Deputy head of 3rd division, UKGB Captain Tornuyev
Approved: Deputy military prosecutor Zavyalov

DECLARATION

Kharkov, November 9, 1937

I, Sergeant Weissband of the Security Service, having studied file no. 9411, re accusations against Weisselberg K. B. and Weissberg A. S., of crimes described in Articles 54-6, 54-11, 54-7 of the Ukrainian Penal Code, find the investigatory material sufficient to prove Weisselberg's counter-revolutionary activity. The investigation of Weisselberg is concluded. I have decided to separate Weisselberg's case from that of Weissberg in readiness for a decision by the court.

Signed: Weissband. Approved: Tornuyev

<u>Extract from Statement No. 13</u>

Decision by People's Commissar of Internal Affairs of the USSR, Commissar General Yezhov, and Attorney General of the USSR, Vyshinsky on October 28, 1937.

We have listened to the evidence against the accused as prepared by Kharkov district NKVD on the orders of the NKVD USSR No. 485, dated August 11,1937.

Decision:

WEISSELBERG K.B.: TO BE SHOT

Signed: Vyshinsky
Extract certified by Yankelovich

CERTIFICATE – Top Secret

The verdict of the NKVD commission and the prosecuting authorities of the USSR, dated October 28, 1937, re Weisselberg K. B., dob 1905, was carried out on December 16, 1937.

Dated: July 4, 1955
Signed: Senior Lieutenant Tikhomirov

PETITION

<u>February 23, 1958</u>

To the Attorney General of the USSR from Anna Davidovna Mykalo, of Tchaikovsky Street, house 8, flat 3, Kharkov.

I earnestly request you to re-examine the case against my husband Konrad Bernadovich Weisselberg, dob 1905, arrested March 5, 1937 in Kharkov and if possible to rehabilitate him. I am convinced that he never committed any crimes against the Soviet Union.

Signed: Mykalo

<u>November 1958</u>

To the head of the Central State Special Archive of the USSR, Moscow.

Due to our re-examination of the Weisselberg case please instruct the Archivist to check and inform us whether there is any evidence of Weisselberg K. B., dob 1905, place of birth, Romania, being a member of the German Intelligence Service.

Signed: Samarkin, head of Investigation Department, Kharkov district.

November 17, 1958 – Reply from Tulyaev:

Among the documents in the USSR Central State Archive there is a secret list, prepared in the Chief Directorate of the German Imperial Chancery, Security Services Division. This list was probably prepared before Germany attacked the Soviet Union in 1941. It mentions Weisselberg Konrad, Doctor of Chemistry, dob October 11, 1905 in Berlad, and currently living in Kharkov.

Weisselberg is listed among those whom the security services consider a serious threat to Germany and who should be arrested at the earliest opportunity.
From documents dated 1939/40. Source: Secret List, Volume No. 1, November 13, 1958.

DECLARATION

There is no compromising material in the case of K. B. Weisselberg in the files of the Kharkov District KGB.

Conclusion of the Investigation, April 17, 1959

I, Captain Gavryushenko, senior investigator of the UKRGB, Kharkov district, having studied the reassessment of case no. 24289 incriminating Weisselberg K. B., dob 1905, etc: find:

Weisselberg K. B. was arrested by the Kharkov district NKVD on March 5, 1937. On October 28, 1937, he was sentenced to death by the NKVD and the Attorney General of the USSR. The verdict was carried out.

According to the charge, his crime consisted of the fact that in 1934 he arrived in the Soviet Union from Austria and formed a criminal relationship with Weissberg, an agent of German intelligence, and became part of Weissberg's revolutionary group. Our re evaluation revealed that Weisselberg was sentenced without justification.

As we see from the case evidence, Weisselberg never admitted any guilt and the entire accusation was built on the vague and dubious testimony of Fomin and Anders who had been arrested on separate charges. Fomin and Anders (aka Kohn) claimed that Weisselberg was a member of Weissberg's counter-revolutionary group.

The documents show that at first Weissberg admitted his guilt, saying that Houtermans recruited him to spy for Germany. But during interrogation in 1939 he retracted his confession. Houtermans, part of the Anders and Fomin investigations, also refused to admit his guilt. Kharkov district KGB has no compromising material on Weisselberg; and according to the Central State Archive, Konrad Weisselberg, PhD, is mentioned in the secret list compiled by the German Security Service as a person representing great danger to fascism and his arrest is recommended. We sought evidence that Weisselberg had been a member of the German Intelligence Service and found none. It is clear that the investigation was conducted by the local NKVD with total disregard for the law. Moreover, neither the final charge, nor the documents of the completed investigation, were shown to Weisselberg.

On all the above grounds I recommend a petition to the court authorities to cancel the decision of the NKVD and the Attorney General of the USSR dated October 20, 1937 in re Weisselberg and to close the file on him.

Signed: Gavryushenko and Samarkin

APPEAL

(An exact repetition of the previous text.)

Signed: Chief Attorney of Kiev Military District, Klimov. 28 February 28, 1959

The tribunal of the Kiev Military District hereby cancels the verdict of October 28, 1937 and concludes the case because there is no charge to answer. Weisselberg K. B. is rehabilitated posthumously.

***

COMMENT

This is the whole of my father's case. I was above all impressed that he consistently denied any guilt. This was very rare. As you know, Alex Weissberg held on for a long time, then admitted his guilt (he took time out, I dare say) and retracted his confession, repeating this pattern several times. And this, although the interrogators treated him more mercifully because he was a foreign citizen.

I can only imagine how they tortured my father. For a long time we kept his underwear (one of the few items he was allowed to change); time after time we found it not specked, but completely soaked with his blood. And this lasted for ten long months.

What is so shocking is not only the groundlessness and absurdity of the charges but the complete lack of any material evidence. Most dreadful of all is that he was given no trial at all. They simply sent Moscow a list of people to be shot, the list was signed by the supreme executioners, Yezhov and Vyshinsky. I can only just imagine my father, a defenseless and a very kind man – according to my mother he wouldn't even hurt a fly – trying to oppose this senseless, angry and merciless Bolshevik system. He stood up for his good name and for his family. He knew what life was like for the family of an enemy of the people. But he never succumbed; he maintained his innocence. This is staggering.

-Alexander Kharlamov

References

[1] Alexander Weissberg-Cybulski, *Hexensabbat*, (Frankfurt, Frankfurter Hefte, 1951, in German), English translation: Alexander Weissberg, *The Accused*, (New York, Simon and Schuster, 1951) and *Conspiracy of Silence*, (Hamish Hamilton, London, 1952). A German edition was released recently in Austria under the title *Im Verhor: Ein Überlebender der Stalinistischen Sauberungen Berichtet*, (Europa Verlag, 1993).

[2] *Physics in a Mad World*, Ed. M. Shifman (World Scientific, 2016).

[3] Yuri Raniuk, *Alexander Weissberg's book "The Accused": Historical Commentaries*, *http://www.sunround.com/club/22/ufti.htm* (in Russian), also in O. Weissberg, *Holodna Gora*, Ed. Yuri Ranyuk, (Prava Ludini, Kharkov, 2010), in Ukrainian.

PLACZEK: CORRESPONDENCE

Chapter 3

Placzek: Correspondence

Placzek did not keep letters addressed to him, nor drafts of those sent to other people. Some letters – not too many, though – survived in archives and with friends to whom they were addressed. This is in sharp contradistinction with one of his close friends and co-authors Rudolf Peierls, whose private archive contained hundreds of letters. Over a thousand of Peierls' letters were catalogued, provided with commentaries and published by Sabine Lee. In the Placzek case we managed to find 42 letters (many of them from the Peierls archive) in addition to those addressed to Charlotte Houtermans and presented in Chapter 1. These letters shed light on events in Placzek's life and help us understand some crucial decisions he made.

Georg Placzek to Chancellor Dr. J.L. Magnes [1]

THE HEBREW UNIVERSITY
Department of Theoretical Physics

Chancellor Dr. J.L. Magnes
Hebrew University of Jerusalem
August 13, 1935

Honorable Mr. Chancellor:

I thankfully confirm to have received your kind letter of August 4. The university's conditions seem to me tantamount to complete rejection of all my suggestions. Thus, as much as I am sorry about that, I cannot take a decision in favor of remaining at the university under the current conditions.

As you wished, I will take the liberty to explain the reasons for my standpoint one more time. The only decisive factor for it is consideration for my scientific work. Since, as follows from your kind letter, there is no prospect for the possibility of experimental work in the foreseeable future, which is important to me, it would be all the more important to me if my suggestions regarding the creation of conditions I deem necessary for successful theoretical work were accepted.

It would delight me very much to be useful to the university now and later, and I would like to ask you to consider me at your disposal in every respect.

With the highest esteem and regard,

Yours faithfully

G. Placzek

∽

---

[1]Courtesy of the Central Archive of The Hebrew University of Jerusalem.

Georg Placzek to Rudolf Peierls [2]

Copenhagen
November 6, 1937

Dear Peierls,

Attached are some pictures for the eternal remembrance of the Copenhagen Conference.[3] You and your wife are much less photogenic, though, than Pauli, Heitler, or Rosenfeld, but nevertheless better than the average Danish movie stars.

Anyway, there is nothing new to report from here. The n-th proof-read version of Bohr's paper is ready every morning. Yet it looks as though Bethe was wrong as regards the 1939 quotation.[4]

Naturally, nothing was going on in Paris either. On the other hand, Beck is very worried because you haven't written to him to say how much you are charging him for the tablecloth, which incidentally his mother-in-law liked very much.

To start off the winter, I try to understand innumerable new radioactive isotopes and isomers that are spat out every fortnight from American cyclotrons to Phys. Rev., but with poor results so far. Please, write me if you know something new about nuclei rotation or non-rotation, etc. Bohr has generally calmed down about the Pauli-Williams effect.

I hope, in the meantime that you have recovered from your illness and that you are already carrying out your new official duties in an exemplary fashion.[5]

With cordial regards to you and your spouse,

Yours,

G. Placzek

∽

---

[2] From Sabine Lee, *Sir Rudolf Peierls: Selected Private And Scientific Correspondence*, (World Scientific, Singapore, 2007), Volume 1, page 541.

[3] Nuclear physics conference in September 1937.

[4] In the last part of their 1937 review paper, Hans Bethe and M. S. Livingston had cited Bohr's joint paper with Fritz Kalckar as published in 1939. (See Rev. Mod. Phys. 9, 382 (1937)).

[5] Rudolf Peierls had taken up his professorship in applied mathematics at Birmingham University in the autumn of 1937.

Georg Placzek to Charlotte Houtermans

B B 41 Vienna 18 14 1800
Date: January 14, 1938
– overnight telegram –

Houtermans
Pension Have
Amaliegade Copenhagen

THE MESSAGES DID NOT ARRIVE HAVE THEM STOPPED AT ONCE WRITE ME A DETAILED LETTER GRANDHOTEL =
PLACZEK

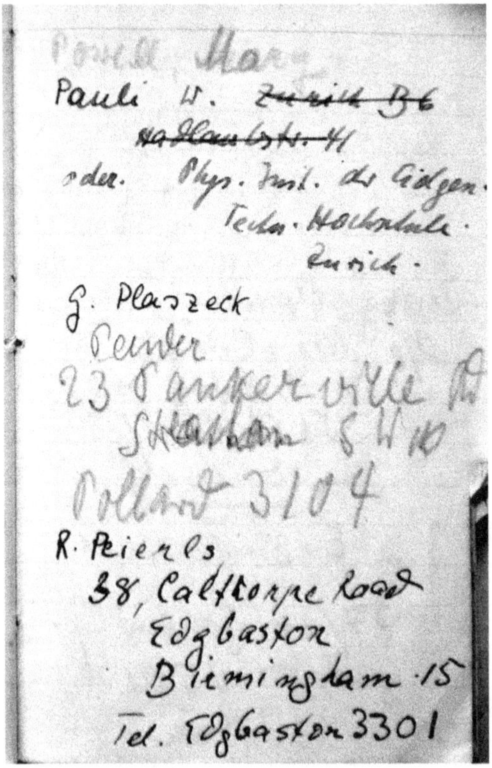

༺

Figure 3.1   A page from Charlotte Houtermans' 1937-39 address book, with Placzek's address in London, 23 Tankerville Road, SW 16, London.

Georg Placzek to Rudolf Peierls[6]

Copenhagen
January 27, 1938

Dear Peierls,

Excuse me for not writing earlier; I've just returned from a prolonged Christmas holiday. Many thanks for the paper, which particularly interests me for several reasons.

a) I hope to find here relief for my guilty conscience; and b) we here are very interested in the questions that emerge when levels are overlapped. Kalckar, who was trying to do something similar, died quite suddenly a few weeks ago of cerebral hemorrhage, as you have probably already heard.

Please let me have some time to read the paper thoroughly. When I am finished with it, I shall write to you in detail and forward the manuscript to Bethe.

There is no other news here. I definitely hope to come to Birmingham, and I am looking forward to it a lot. At the moment I cannot leave, or else Bohr will be orphaned, but I hope I'll be able to in early spring.

Recently Frau Houtermans showed up here with her children. She left her husband at the place where the majority of his Russian colleagues are now staying. She herself had a fairly easy time of it and thanks to a series of lucky circumstances they even managed to get her safe passage to London.

Okay, I'll write again soon. Best regards to you and your wife.

Yours, G. Placzek

P. S. Just now Bohr tells me he's received a letter from you; I'll see to it that you [get] an answer soon.

∾

---

[6]From Sabine Lee, *Sir Rudolf Peierls: Selected Private And Scientific Correspondence*, (World Scientific, Singapore, 2007), Volume 1, page 550.

Rudolf Peierls to Georg Placzek[7]

Birmingham
April 22, [19]38

Dear Placzek,

Attached are, first of all, the manuscripts of [Guido] Beck[8] and [André] Mercier,[9] which I naturally forgot to give you. Additionally, a copy of a card from Frau Schrö to acquaintances in England. This is only for your information.

I talked to Schnax[10] by telephone, and she seemed very disappointed that Bohr does not want to write. She realized, of course, that a letter to the president of the academy makes no sense, but she had obviously hoped that he would write directly to higher authorities. I didn't quite know how to answer her question as to why he wouldn't do so.[11]

Today I received a letter from Bethe, where he writes, among other things: "Weisskopf and I are primarily interested in the photoeffect. Incidentally, Placzek claims that you told him that the paper by Kalckar, Oppenheimer, and Serber on the photoeffect was fine.[12] He (or you) can't be serious, though. I found the paper by Bohr in *Nature* on this topic to be regrettable: not only are the experimental results not as he says, but also the "explanation" seems reactionary and senseless to me.[13] Why, he now even takes the crystal model more seriously than anyone ever took the Hartree model!"

I repeat this only in order to help clarify the situation. In particular, I believe it follows that Bethe and Weisskopf probably know why the Kalckar-Oppenheimer equation is wrong, and pretty much have the correct one. In addition, Bethe told me that on July 18 he will leave for Europe and return around September 18.

---

[7]From Sabine Lee, *Sir Rudolf Peierls: Selected Private And Scientific Correspondence*, (World Scientific, Singapore, 2007), Volume 1, page 565.

[8]Guido Beck (1903-1988) was a physicist born in in the Kingdom of Bohemia (Austria-Hungary), of Jewish descent. He received his doctorate in 1925 in Vienna, under Hans Thirring. In 1935-1937 he taught theoretical physics at Odessa University in the USSR.

[9]André Mercier (1913-1999) was a Swiss physicist, professor at the University of Bern, Dean of the Faculty of Sciences and later President of the University of Bern, Switzerland. In 1971 he was the main founder and secretary to the International Committee on General Relativity and Gravitation.

[10]Schnax was the nickname of Charlotte Houtermans.

[11] Charlotte Houtermans had hoped for Bohr's intervention on behalf of her husband arrested by Soviet police.

[12] F. Kalckar, R. Oppenheimer, and R. Serber, "Note on Nuclear Photoeffect at High Energies," *Phys. Rev.* **52**, 273-78 (1937).

[13]N. Bohr, "Nuclear Photo-Effects," *Nature*, **141**, 326-27 (1938).

Concerning physics, I have no news. Regarding the contribution to states with energies beyond the width, the more I think it over, the more I am convinced that they cannot matter. E. g., in the case of the light absorption by macroscopic bodies, where the situation is understood rather well (or at least is believed to be so), they make no difference. But it is naturally not easy to prove this convincingly, without long x-ing, via expansions in series, in non-orthogonal function systems, or in similar ways.

With cordial regards, including to Frau Bensch, and once more "tak før sidst." [Danish for Thank you – literally "Thanks for the last time."]

Yours [R. E. Peierls]

Georg Placzek to Rudolf Peierls[14]

Copenhagen
April 27, 1938

Dear Peierls,

Thank you very much for your letter. I have delivered the documents, so that they can no longer sit around quietly getting moldy. There is no particular advance in my science – in the field of gamma – it is temporarily at a standstill. In the meanwhile I am waiting for a delivery from you of the war materials that I asked you for in my last letter, in order to proceed.[15]

Now, Bohr has compiled a note on photo-effect for *Nature*,[16] He meant that he would make everything clear. But as it seems to me, one cannot gather anything from this paper. It seems to me however, that after it appears no one will understand it anyway. Therefore I have dedicated all my energy for aborting the publication plan; for right now I have succeeded in having it postponed. If this continues until Greek Calends this will be all right.

I was not in the least surprised by Bethe's arguments on photo-effect; anyone could deduce that, if the selectivity doesn't exist and the O.K. equation[17] is wrong, so Weisskopf's equation is not right, too. It is completely obscure for me from where the idea comes that I stated the validity of the O.K. equation at high energies as you have cited this. In the initial letter, I only stated (and with emphasis) that Weisskopf's equation was derived by Abyssinian methods, and therefore I don't believe it, and that I share this judgment with you.

The false statement that I made is in the last letter to Weisskopf which Bethe would not have had access to and even there I never stated that the O.-K. formula for low photon energies (long wavelengths) is valid – this certainly would be a great nonsense. In addition I haven't written to Weisskopf yet, because Bohr does not want to update him until we are in agreement. I won't wait any longer though because I

---

[14] From Sabine Lee, *Sir Rudolf Peierls: Selected Private And Scientific Correspondence*, (World Scientific, Singapore, 2007), Volume 1, page 568.

[15] Peierls had written another letter on April 26, 1938. A carbon copy exists in Bodleian, but due to the large number of missing handwritten passages its content is unclear. Peierls Papers, MS. Eng. Misc. b203, C. 30.

[16] N. Bohr, "Nuclear Photo-Effects," *Nature*, **141**, 326-327 (1938).

[17] Presumably, the Kalckar-Oppenheimer-Serber equation, see footnote on page 152.

can't. Please write to Bethe on your own.

[Christian] Møller, [Léon] Rosenfeld, and I have all received long letters from Mrs. Houtermans.[18] Bohr will consider again if he should write. I will update you on the decision. Other then that there is currently nothing new. Heartfelt greetings,

Yours

Georg Placzek

P. S. Thanks for the card from Frau S. Its content is definite. The trip to Berlin included Planck-Fest.[19]

P. S. 2 The manuscript is being sent separately.

P. S. 3 Regards from Bohr who thanks for your letter.[20] The division passed well. He will write then.

∽

---

[18]Niels Bohr did not intervene on this occasion. In contrast, he tried to act on Lev Landau's behalf.

[19]On April 23, 1938, Max Planck celebrated his 80th birthday.

[20]Letter from Peierls to Niels Bohr, April 22, 1938. Peierls Papers, MS. Eng. Misc. b203, C. 30.

Georg Placzek to Rudolf Peierls[21]

May 16, 1938

Dear Peierls,

[...] Now, on practical matters. I have received Bohr's telegram. In general, we believed that it would be probably OK if we could finally prepare the study all together. Therefore Bohr sent a telegram to Weisskopf for him to arrange this from the American side when I arrive with Queen Mary (from June, 1) instead of Aquitania (from May, 25). Subsequent steps will depend on Weisskopf's answer.

1. <u>He supports.</u> On Friday 20th, I leave for Prague, since I already cannot cancel this agreement, and on Sunday 22nd, evening, I will be back in Copenhagen, where I stay till 28th, and go to Paris from there.

2. <u>He refuses.</u> In this case, <u>I</u> must go from Prague on Saturday evening to Paris. <u>Bohr</u> in this case will consider if he would ask you to shift your visit, since just this week he has to deal with Rosenfeld (less work will remain there, if you and me could work together as in Case 1), and if you eventually (depending on circumstances) will visit England later. <u>I</u> will gladly see you in this case under any conditions, wherefore two ways are possible. Either you and your spouse have a desire for a short tour to Paris, where we could meet each other on Wednesday, or (if the above is inconvenient for you) I can fly from Prague to London on Saturday evening and stay till Sunday evening, and we, if it is convenient for you, can settle all open questions.

In Case 1, it would be appropriate if you first leave on Saturday 21st, and on Sunday evening come back, since otherwise you must take quick decision, and the answer from Weisskopf will hardly come before Wednesday. We shall wire it to you at once.

Excuse me, please, this pseudodramatic disorder, but I cannot say more.

Yours G. Placzek

○❧

---

[21]From Sabine Lee, *Sir Rudolf Peierls: Selected Private And Scientific Correspondence*, (World Scientific, Singapore, 2007), Volume 1, page 578-579.

Georg Placzek to Rudolf Peierls[22]

Perro Caliente[23]
July 17, 1938

Dear Peierls,

[...] Concerning American relations, you will have Bethe as physicist [unclear -MS], who will probably appear at about one time with this letter, by verbal information. We, Weisskopf and I, reached hitherto with no dramatic events to here, in the wilderness of New Mexico, where Oppenheimer's ranch is located, and we even have converted Oppenheimer to the truth in photoeffect topic. (Weisskopf, though, has long ago modified his Abyssinian method, so that it gives right results, but of course, after studying your paper, he attaches to it no significance).

Besides, on these days I found a book, which can more than compensate any private journey around America and makes visiting this continent superfluous: "Odnoetazhnaya America" ("One-Storied America") by Ilf and Petrov[24] (the authors of "Zolotoy Telenok" – "Little Golden Calf"). I recommend the book to you enthusiastically.

Have you any authentic communications[25] about Landau's fate? Please, write me: c/o Bloch, Stanford University (California).

I hope, you enjoy your vacation despite Buch and Bohr's B. A. lecture. Heartfelt greetings,

Yours,
G. Placzek

∞

---

[22] From Sabine Lee, *Sir Rudolf Peierls: Selected Private And Scientific Correspondence*, (World Scientific, Singapore, 2007), Volume 1, page 501 502.

[23] Oppenheimer's ranch in Pecos Valley near Los Alamos, NM.

[24] Ilya Ilf and Evgeny Petrov were Soviet humorists active at the end of the 1920s and in the 1930s. *Little Golden Calf* was a rollicking picaresque novel of farcical adventures within a framework of telling satire on Soviet life in the late 1920s. In 1936, following a tour of the United States, Ilf and Petrov published "One-Storied America," a witty account of their automobile trip across the US written in the form of an exposé of the materialistic and uncultured character of American life.

[25] Lev Landau was arrested by the NKVD in Moscow on April 28, 1938.

From The Bethe-Peierls Correspondence [26]

Peierls to Bethe

Birmingham, August 24, [19]38 [27]

[...] Another piece of news is that Jones, who was my successor at the Mond Laboratory, got a readership at the Imperial College. Cockroft is therefore looking for someone new. I mentioned Placzek, and he seemed to be very excited by the idea. We therefore decided to telegraph him to ask whether he is still free, as I have no idea how his negotiations with Cornell are going. I wanted to tell you, in any case. If he hasn't definitely decided what to do with Cornell, and if this new proposition has any influence on his decision regarding Cornell, I presume that he will immediately contact you. I will keep you informed about new developments in Cambridge in any case. Also, please tell me when you are coming and which ship you will take.

With best greetings from all of us, and also to your mother.

Yours, [Rudi]

Peierls to Bethe

Birmingham, August 24, [19]38 [28]

[...] I haven't received an answer from Placzek in response to the telegram sent on Tuesday. Do you know anything about where he is, and whether we can expect that a telegram addressed to "c/o Bloch, Physics Department, Stanford University" would be forwarded to him?

Bethe to Peierls

Seelisberg, August 27, [19]38 [29]

Dear Peierls,

This is all very, very interesting, especially your news of Placzek. I am very happy and relieved about this. Personally, I would of course have liked to have him at Cornell, but − 1. Cornell would have been totally wrong for him as one theorist is really enough for any institute. In this regard, Cambridge would be perfect as it urgently needs a real theorist. 2. I actually need a slave and Placzek does too; in Cornell,

---

[26] August-September 1938, in Sabine Lee, *The Bethe-Peierls Correspondence*, (World Scientific, Singapore, 2007).

[27] Page 238. At that time − August 1938 − Placzek was on a tour in the United States, with Weisskopf.

[28] Page 241.

[29] Page 244.

however, there wouldn't be any more jobs for slaves. 3. There would be difficulties with the other people at our institute (an improved version of the Gibbs-Richtmeyer conflict [30]), and one shouldn't make his political situation even worse. 4. Placzek would probably feel more comfortable in Europe than in a small town in America. 5. As far as I know the salary is the same ($2000 vs. £400), but in Cambridge it's easier to earn some extra money and the salaries of other jobs are no higher. 6. Cornell was just meant as a stepping-stone to a better job in America, but it is doubtful that the Americans would be reasonable enough to accept Placzek's "un-American" qualities.[31] By the way, is the Mond position for five years?

So all in all, I would prefer it if P[laczek] went to Cambridge. The only thing is, would Ewald's chances at Cambridge be lowered because of another foreigner? But actually I think that the Mond and Cavendish, and especially nuclei and crystals, are linearly independent. Does Bragg have a say in things, there at Mond?

I'm not just telling you all these reasons you already know to reduce the post office's profit. It would probably be a good idea to tell them to Placzek himself. It wouldn't be very nice if I did this, but it would be much nicer if, for example, you explained to him the advantages of Cambridge as compared with Cornell. Would that be possible? I think this could influence his decision.

About a week ago, I received a telegram from Placzek in which he told me that he would like to come to Cornell but just for the second semester.[32] In the first semester, he would definitely like to stay in Europe. I telegraphed him that this would be acceptable, and telegraphed the same to Gibbs. I haven't received an answer yet. Cornell can be cancelled at any time, of course, especially as we have an applicant (Rose) who is ready to come whenever we want. By the way, I would take Rose for one more year, then either Critchfield (from Teller) or Schwinger (from Rabi), if Placzek doesn't come. It is therefore possible to reserve Cornell for Placzek at least until February [1939], if Cambridge doesn't come through. How are things going in Cambridge,

---

[30]R. Clifton Gibbs and Floyd K. Richtmeyer, both professors at Cornell, were engaged in a continual feud, and as Bethe had been recruited by the Head of Department, Gibbs, his standing with Richtmeyer was on shaky grounds. -SL

[31]Perhaps, Bethe hints to leftist political persuasions of Placzek. As we know, they completely faded away after Placzek's visit to the USSR in 1936-37.

[32]Placzek planned to spend three months, from October till December 1938, in Paris, with Joliot-Curie and von Halban.

by the way? Does it only depend on Placzek's decision, or does anyone else need to be asked? [...]

Yours, Hans

Peierls to Bethe

Birmingham, August 29, 1938 [33]

Thank you for your letter. I have now received Placzek's answer. In principle, he is very interested, but he told me that he had to spend the spring term in Cornell and that he couldn't decide right now as negotiations with Paris were still going on. He therefore asks for a few weeks to think it over. I immediately forwarded this message to Cockroft, of course, and I actually hope that Cockroft will wait for him. It is my impression that everything is all right in Cambridge. In principle Bragg has something to say about it, but I think he is sufficiently unpopular in the Cavendish itself and therefore won't try to interfere with matters at the Mond unnecessarily. It was also especially fortunate that Bohr was in Cambridge, as Cockroft immediately sent Bohr to speak with Bragg, which seems to have had a good effect. It will take a long time before the whole thing is officially decided, of course, but I'm not too worried about that. If you could keep him on the back burner at Cornell for a while, it would certainly be very generous. Officially the job is for one year, but that means nothing. In the beginning there was grant money for five years; three are already passed, and I don't know whether the grant will be renewed. But I'm guessing that it will be renewed, as without a theorist [...] the Mond laboratory could hardly exist. Anyway, it seems certain that once Placzek comes here, provided that he doesn't become too unpopular (which seems to be much more difficult in England than in America, and his manners aren't as bad as Moritz's), he will be able to stay permanently, because Cambridge lacks a theorist.

I am especially pleased that you don't blame "us Englishmen" for trying to snatch Placzek. I think at the moment it is *not* necessary to forward your five points to Placzek, as his telegram seems to show that he has understood all by himself that Cambridge would be better for him, and that he just wants to keep his promise to Cornell for the spring term. I will therefore carefully tell Placzek that I have the impression that you wouldn't blame him for cancelling Cornell, if it improved his situation at Cambridge. This is actually something you could tell him

---

[33] Page 248.

too, of course.

I will also tell Cockroft – with a much more cautious formulation – that it isn't impossible that Placzek would come to Cambridge for the whole winter if he really wanted him to do so, and anyway that he shouldn't take offense. I don't think that this whole situation has anything to do with Ewald's chances. [...]

Sincerely yours, Rudi

Bethe to Peierls

Seelisberg, September 6, [19]38 [34]

Dear Rudi,

[...] First, thank you for your information about Placzek. I have now received a letter from him, written before he received your offer, when he didn't know about either Cornell or Paris. If he got Paris as well as Cornell, he was hoping to try each one for half a year. I have now sent him a letter, which was easy because of his positive opinion of Cambridge. I wrote that I had heard from you etc., and that I would like him to come to Cornell for half a year but that he should cancel it if Cambridge appeared to rely on his coming earlier in any way. I also hinted at my 5 reasons... [and asked him] to come to Cornell at the end of September for a couple of days in any case, to talk about all this. I also sent a letter to Gibbs, the head of my department, to tell him about the great challenges waiting for Placzek in Cambridge, and that we would have to do without him if Cambridge asked for him, and that we should be grateful to get him if for any reason he decided to come back here a year later. I think that everything is all right here, and that Cornell is still a possibility for Placzek until Cambridge decides officially (which presumably can be expected before 1939?) [...]

Best wishes to the whole family, Hans

Bethe to Peierls

[Ithaca], December 12, 1938 [35]

[...] I don't understand Placzek's decisions. [...].

∽ ●

---

[34] Page 252.
[35] Page 266.

Rudolf Peierls to Georg Placzek[36]

Birmingham
August 27, 1938

Dear Placzek,

Recently I received your wire. I communicated its content to Cockroft, and of course it depends on him – whether he would wait a couple of weeks or take any further steps meanwhile. Probably, you will already have a wire notice on this issue before receiving this letter.

On the history of this matter: the point is that Jones, my interim successor in Cambridge, got a readership in London and departs there by October 1st. As it seems, Cockroft liked much the idea to obtain you for this position, and then in the Brit[ish] Ass[ociation] were both Bohr and Bragg; so the matter was settled unexpectedly easy. Of course, it is not an official decision yet – to offer the position to you, as you can well understand from the facts, but I think it is a pure formality. Without saying, Bragg is much less reliable, than Rutherford was, but it is utterly improbable that he would trip up in the last moment. Practically, everything depends on how Cockroft will hasten. My impression is that he tries to wait for you: he seems not to be delighted by other available competitors for this position.

As you know well, the very position is purely consultative. The concept is to be slightly curious about things that people around do. (This gives much pleasure, though, for there are very interesting and quite unexpected results about liquid helium, etc.) Of course, it is not required to produce at once a theory for each experiment, but mostly people need something about phenomenological or thermodynamical interrelations between their different experiments to be revealed, explained, and in general, a theory proposed on their issues. They need to be told where they may be wrong. Further, questions of possible measuring means, estimates of experimental uncertainties, evaluation of results, etc. should be discussed. In these issues it is not necessary to make decisions for all, but rather explain, how it should be done, and wherefrom it follows, if you can.

Beyond the above, one must (informally), of course, contact with Cavendish, and I can only imagine, how many useful initiatives you would propose, though no explanations are needed for you. The salary,

---

[36]From Sabine Lee, *Sir Rudolf Peierls: Selected Private And Scientific Correspondence*, (World Scientific, Singapore, 2007), Volume 1, pages 610-611.

as you know, is £400 minus 5% for superannuation. But in Cambridge you may earn *á la carte*. If you are invited to a faculty to read lectures, for a three-hour lecture during a term you will obtain £50 more (while you have asked for yourself only 15). Then, it is easy to obtain supervision, and one hour per week for a term adds £3 – 3 – 0 (for several students taken at a time, respectively more). But all these matters require a long speedup. For lectures, one should be a member of the University, and for supervision one should be attached to a college. For the first winter you probably cannot converge here.

This is for your information. If I hear something definite from Cockroft, namely if he will finally decide, to wait or not to wait, I shall let you know. I go to Copenhagen on August 31st for a fortnight, and then return. (Naturally, I have an idea to write the paper.)

With cordial regards,

Yours,
R. E. Peierls

Georg Placzek to Rudolf Peierls [37]

Pasadena
September 4, 1938

Dear Peierls,

[...] About business. First of all, may I ask you, as blessed Bukharin [38] asked me (when once I, so to say, personally represented international science and solicited for Landau, trying to convince Bukharin that they should now and then let him travel abroad), namely: is your *démarche* official, officious, or unofficial?

Further, I would like to enquire about the current situation at Cambridge. Is it true that they have decided, against all expectations, to revive Nuclear Physics and assign the lunar position to a nuclear physicist? Or is it the case, as feared, that they want nuclear physics to wither away slowly and, independent of this (e.g. by Bohr's solicitation), kindly offered me the position as a sort of sinecure?

My situation is as follows: Bethe certainly told you about the Cornell position, that he proposed it to me. It is not quite appropriate, but if I wanted to remain in America it would be just a reasonable beginning, as they said.

Before receiving your cable, I accepted Cornell for a start, for one semester until February. The Paris affair is at a standstill, I have heard nothing from Paris officially, although Joliot had promised his answer in July. Privately I heard from Halban [39] that, due to the absence of some necessary people, the meeting at which my appointment was supposed to be decided was shifted from July to October.

Now I can put the Parisians out of my head, since I hope to accept Cambridge without waiting for Collège de France's decision (as for

---

[37] From Sabine Lee, *Sir Rudolf Peierls: Selected Private And Scientific Correspondence*, (World Scientific, Singapore, 2007), Volume 1, pages 624-625.

[38] Nikolai Bukharin (1888-1938) was a Russian Bolshevik revolutionary, Soviet politician and prolific author on revolutionary theory. On Stalin's order he was arrested by the NKVD on February 27, 1937, and then sentenced to death by a staged "public" trial. He was executed by firing squad on March 15, 1938.

[39] Hans Heinrich von Halban (1908-1964) was a French physicist of Austrian-Jewish descent. In 1937 Halban was invited to join the team of Frédéric Joliot-Curie at the Collège de France in Paris. With the German occupation of Paris in May 1940, Halban and Kowarski left Paris for England with the supply of heavy water, radium and their research documentation, as instructed by Joliot-Curie. Later he was sent by Churchill's government to Montreal as head of the research laboratories at the Montreal Laboratory (part of the Manhattan Project.) George Placzek worked for this laboratory too. Moreover, he married Halban's ex-wife Els, née Andriesse.

Cornell, so far I am bound there for one semester only, and in any case it makes no harm to Paris).

I have asked you to maximally speed up making the decision. The question is, for how long the issue could be under consideration at Cambridge? In case you have not yet written anything to me, please, wire to Stanford. My locations in following days are not certain, but my mail will be forwarded from Stanford to me. Please, wire me here.

The address is: c/o Rabi, Physics Dept., Columbia Univ., New York City. I think, on October 5 I will leave New York with *Queen Mary*. If it is absolutely necessary, I can leave 14 days earlier, of course, on my own expense – in this case I will need a prompt notice by wire.

Now I must finish in order to send the letter in time. Many thanks and best regards,

Yours
G. Placzek

∽

Georg Placzek to Rudolf Peierls[40]

Paris, October 17, [1938]

Dear Peierls,

Many thanks for your card. After long torments, I finally gave birth to a letter to Cockroft. I wrote to him that in view of the positive decision about Paris I will be busy all year, but that if he finds it difficult to fill the position immediately and thinks that I would be useful helping out on a temporary basis, I would be happy to try to find an arrangement that would permit me to be at his disposal for part of the coming months. I hope this is regarded as [mere] politeness, as intended, for if it were ever to get to that point I'm not sure it could really be organized here. I did not write further expressions of courtesy to him (such as how glad I would have been to work with him and how much I appreciate the honor of a position in Cambridge, etc.) because of my poor command of English. If you meet him any time soon and could tell him something to that effect, it might be a good idea.

There's nothing much happening here; you can imagine the general political mood. I'd like to see you soon. I don't know how long will it take to get my documents in order and to be mobile again. Please consider if and when you can come over for a weekend, which would be really splendid.

I hear that Shoenberg is supposed to be in Cambridge now; do you know anything reliable about [Lan]dau?

In America Wigner is spreading horror stories about your system of functions, promoting in particular the opinion that it is not complete. It would therefore be a good idea if you were to publish your argument to the contrary. I regret that I did not meet Wigner himself.

The Washingtonians want to try and invite you for their next conference on deep freezing, but they do not know if they have the money yet. I tell you this under the deep seal of secrecy.

My best regards to you and your wife,

Yours,
G. P. m. p.

∾

---

[40]From Sabine Lee, *Sir Rudolf Peierls: Selected Private And Scientific Correspondence*, (World Scientific, Singapore, 2007), Volume 1, page 630-631.

Rudolf Peierls to Georg Placzek[41]

Birmingham, October 22, [19]38

Dear Placzek,

Many thanks for your letter. I shall now arrange things so that next week "by chance" I will show up Cambridge. Then I can first explain the situation to Cockroft in the sense that you wish, if necessary, and also try to advise him to take Wick, who, as I suppose, would be quite glad to come, and who is presumably the best person for this position now.

I spoke to Shoenberg. He had no more to tell about [Lan]dau than we already knew (or feared). Rumer[42] and Hellmann[43] belong to the same group. [Walter] Zehden is running around here in England; he got here via Berlin but had to leave his Russian wife and child in M[oscow], and hasn't corresponded with her for months. What a world we live in.

I would like to see you very much also, but I don't know yet when I'll be able to take a jaunt to Paris. I haven't heard anything from Bohr for some time. He wrote me at the end of September that he has now sent a letter concerning [Lan]dau he's been planning for a long time, in the appropriate form. But apart from that he probably has a guilty conscience, as he hasn't touched the paper once since my departure. I'm beginning to lose hope that it will ever come to anything. By the way, have you found your calculations with the Pointing vector in the meantime? I've always been too lazy to do routine calculations,[44] but when Bohr finally comes to a conclusion, it might be very nice to write down the correct factors.

I'd rather not write about the political situation. It's just too annoying.

---

[41] From Sabine Lee, *Sir Rudolf Peierls: Selected Private And Scientific Correspondence*, (World Scientific, Singapore, 2007), Volume 1, page 632-633.

[42] Yuri Borisovich Rumer (1901-1985) was a Soviet physicist known in the West as Georg Rumer. At the early stages of his career he worked as an assistant of Max Born at the University of Göttingen in Germany. Rumer was arrested in 1938 in Moscow as an accomplice of the "enemy of the people Landau" when he was going with friends to celebrate his birthday.

[43] Hans Hellmann (1903 1938) was a German physicist and chemist. After the Nazi rise to power, Hellmann immigrated to the Soviet Union, taking up a position in Moscow. During the Great Purge, he was arrested on May 10, 1938, and executed on May 29 that same year. See M. Shifman, *Physics in a Mad World*, (World Scientific, Singapore, 2015): Corrections and Addenda, on www.academia.edu, pages 1-4.

[44] In German original: "Ich war noch immer zu faul, das *durchzuixen*..." Infinitive form of the verb is "durchixen." Colloquial meaning "do something mechanically." Originates from being x overwritten by typewriter.

Best regards,

Yours

[R. E. Peierls]

P.S. By the way, what is m. p.? I understand, M. P., or m. p.? Emerited professor? merchant of physics? miraculous prodigy? meticulously polite? Oh, no, I cannot guess it.

∽

Niels Bohr to Rudolf Peierls[45]

Copenhagen, June 6, 1939

Dear Peierls,

I thank you for your kind letter and need not to say how happy I am to know – as I first heard from Cockroft a few days ago – that Landau is now back in Kapitza's Institute and has taken up work again. I look forward very much to come to Birmingham and it shall certainly be a great pleasure to me to accept the kind invitation of Mrs. Peierls and you to stay in your home where I spent so delightful a day on my last visit. I have been very thankful for your kind interest in my little article on the human cultures [46] and I hope that you were fairly satisfied with the final form of the article when it appeared in *Nature*. On my next visit I should be very happy if we could work in [for] a few days together on a brief note to *Nature* about the main results of our common work on the nuclear dispersion theory and, if it would suit you, I shall be glad to come to Birmingham about the 28th of June. Due to the unquiet times and my unavoidable occupation with the fission problem I regret that I have not yet found opportunity to finish our article with Placzek, but, as Placzek suggested, it would be nice if a short account of the result could appear in *Nature* in a near future, and I shall bring with me a draft Placzek and I have written.[47] When we meet I hope also that we can arrange to work together at some later quiet time to finish the greater article to appear in the Proceedings of the Danish Academy. Perhaps you could come to Denmark for this

---

[45]Sabine Lee, Vol. 1, page 670.

[46]N. Bohr, "Natural Philosophy and Human Cultures," *Nature* 143, (1939), 268-72.

[47]This summary note did, in fact, appear: N. Bohr, R. Peierls, G. Placzek, "Nuclear Reactions in the Continuous Energy Region," *Nature*, 144, 200-201 (1939). -SL

purpose some time in August or September, but about that we can best speak when we meet in Birmingham.[48]

With kindest regards and best wishes to Mrs. Peierls and yourself from my wife and Yours,

Niels Bohr

∽

Rudolf Peierls to Niels Bohr[49]

Birmingham, September 13, 1939

Dear Professor Bohr,

Here is, at last, the revised manuscript of our paper. I have completely rewritten the introduction on the lines of our discussion. I found that after that, the other sections could be considerably shortened, but apart from this they are essentially unchanged. I have omitted from the introduction the part on the limitation of the picture for very high energies, which appears now irrelevant, (but which could also easily be inserted) and I have omitted the quantitative discussion of the photo-effect, since I thought one could not add in this respect to what has been said in the note to *Nature*.

I am sending a copy to Placzek. I hope that you will now find the paper readable, but if you do not, and if it will be difficult to continue the correspondence, I agree, of course, to any alterations which you care to make.

With best wishes and kindest regards to you and Mrs. Bohr from all of us,

Yours very sincerely,

[Rudolf Peierls]

∽

---

[48]Collaboration on the project was interrupted by the war, see Peierls' letter to Bohr of February 6, 1948, page 190.
[49]Sabine Lee, Vol. 1, p. 685.

Georg Placzek to Hans Bethe[50]

New York
July 31, 1942

Dear Bethes:

Many thanks for the letters and apologies for the delay in writing, which this time has better reasons than usual. I have been through a rather hectic time these days. We all know, of course, how hard it is to get into this country, but I have now learned that this is chicken feed compared to the effort needed to get out of it. Anyhow, everything is settled now; in record time I believe, and now I am all set. If after the war I ever should be forced to write up my Paris lectures, I guess I shall add an appendix about the inverse process. But lets hope there will be less demand for it than for the original thing.

Last week I was in Ithaca for a day, evacuating all of my desks completely (!). And all I left in my office is a few reprints which by future historians will probably be classed as part of old Nichols[51] remains. I am curious to know whether my ghost will behave as aggressively as his.

I am sorry I was unable to carry out your directions in re Churchill. The old boy was guarded by an hysterical old hag, Miss Dean of Ithaca, who prevented my access to him. So instead of filling up your tank I registered you with the gasoline rationing board. Basic rationing book A is enclosed herewith. This is what everybody gets according to present procedure. Application for supplementary rations has to be made later. Certainly the book will enable you to fill up the tank and use the car right away when you come to Ithaca, The car license is in the hands of George Winter.

Ithaca looks rather deserted now as you can imagine and old Gibbs[52] looks like Dürer's etching "Melancholia." He was exceedingly nice to me and I felt great pity for him but this did not prevent me from establishing a short-time record for our final goodbye conference. Four minutes, believe it or not. Witnessed by Weisskopf who had foreseen two hours as a minimum.

As you can imagine, I am envying you very much for being in

---

[50]Courtesy of the AIP Niels Bohr Library.

[51]Apparently, George Placzek means Edward Leamington Nichols who was a professor at Cornell University and died in 1937.

[52]Roswell Clifton Gibbs (1878-1966) was Chairman of the Department of Physics at Cornell University from 1934 to 1946.

California. Even if you have to stay in the Durant (hope you have found something better in the meantime), but as the man in the train in St. Louis (story number ?) has found out already, one cannot go West and East at the same time.

I have to finish now. So please keep the stories numbered and the country going and greet all the friends. And thanks for having kept me alive all these years. Hope we will hear from each other soon. All good wishes,

Yours, G. P.

ლ

Commentary: From this letter one can infer that Placzek was planning to travel abroad in the summer of 1942. Indeed, Placzek's Los Alamos CV (see page 237) shows that in 1942 he spent four months at the Cavendish Laboratory in England – from August till November. From the same CV we infer that he worked there on chain reactions. From other sources it is known that Placzek arrived in Montreal as a member of the British Mission in late December 1942.

Figure 3.2   Dürer's "Melancholia."

Placzek to Hans von Halban[53]

Montreal
January 14, 1943

NOTE TO DR. HALBAN [54]
RE: PERSONNEL T.P.D

I should like to have your reaction to the following: Dr. W. Lamb[55] – theoretical physicist, pupil of Oppenheimer – presently instructor or assistant – several papers on slow neutrons, – professor at Columbia; is so far not fully employed in the American war effort because his wife is an enemy alien (German). Do you think the man is of interest for us in principle, or not? If the answer is in the affirmative, I should like to find out more particulars tomorrow in New York.

G. Placzek

∽

Commentary: From Chapter 1 (see page 42) we know that the first of the staff of Montreal Laboratory – P. Auger, G. Placzek, H. Halban, and a few others – arrived at about the end of the year 1942. Temporarily they occupied an old residence at 3470 Simpson street belonging to McGill University, but three months later moved to a large new building of the University of Montreal. As is seen from the letter above and Chapter 1, already in January 1943 George Placzek was preoccupied with recruiting talented physicists for the theoretical group under his leadership. On February 9, 1943, he prepared for higher authorities a detailed note entitled *On Personal Situation of Theoretical Physics Division*. One can find a reprint of this note in M.M.R. Williams, *Progress in Nuclear Energy*, **36**, 239 (2000). M. Williams notes: "The recruitment of first rate scientists to the Atomic Energy Project, and their retention, was by no means easy. At times, especially when there were difficulties with the United States, morale would drop and some staff members would look 'South' for more interesting work."

---

[53]Courtesy of The Public Record Office, Kew, UK. Classification AB 1/144.

[54]Hans von Halban (1908-1964) was a French physicist, of Austrian-Jewish descent. In 1942, Halban was sent to Montreal as head of the research laboratories at the Montreal Laboratory, part of the nascent Manhattan Project.

[55]Willis Eugene Lamb Jr. (1913-2008) won the Nobel Prize in Physics in 1955 "for his discoveries concerning the fine structure of the hydrogen spectrum," the so-called Lamb shift.

Placzek to Bohr[56]

NATIONAL RESEARCH COUNCIL
CANADA
(MONTREAL LABORATORY)
P.O. Box 159, Station H, MONTREAL, P.Q.,
October 21, 1943

Professor Niels Bohr,
c/o W. A. Akers, Esq.,
London

Dear Professor Bohr,

The good news of your and your family's safe arrival in England[57] has, of course, made us all here extremely happy. During all these years I have often been thinking of our last discussion in Princeton in the beginning of 1939, and of how right you then were not only about 235, but also in your political optimism about the future. Much has happened since then and I hardly can await the day when it will be possible for me to see you again. I very much hope it will not be too long now before you visit this side of the Atlantic. You have probably heard in detail from your English friends who have recently been over here what the various people here and in the United States are doing. I expect you will be too busy now to write soon, but I hope to hear very soon from Peierls how you are, what your plans are and also what he might have heard from you about the fate of our friends in Denmark.

Yours,

G. Placzek.

⌘

---

[56] Courtesy of The Public Record Office, Kew, UK.

[57] In September of 1943, when Bohr's arrest in occupied Denmark became imminent, he fled Denmark. Bohr reached London via Sweden.

H. von Halban to W. A. Akers[58]

MOST SECRET

NATIONAL RESEARCH COUNCIL

CANADA

(MONTREAL LABORATORY)

P.O. Box 159, Station H, MONTREAL, P.Q.,

March 29, 1944

W. A. Akers, Esq.,
Department of Scientific and Industrial Research,
Directorate of Tube Alloys,
16 Old Queen St.,
London, S.W.1.

Dear Akers,

Please excuse my not thanking you for your kind letter of March 7th. You will understand that since I was forced to send my two cables of March 17th in spite of this letter from you I wanted to delay sending you a written reply.

You can imagine that we were very pleased when the High Commissioner visited us yesterday to explain to us the intentions of the Chancellor in case we should not succeed in obtaining collaboration with the United States in the near future. The facts he reported in themselves as well as his very frank way of dealing with everybody, meant a terrific encouragement for all. In agreement with the High Commissioner we had assembled for this meeting all United Kingdom personnel with General Clearance plus Freundlich, that is, Auger, Jackson, Newell, Paneth, Anderson, Bauer, Ginns, Greenwood, Arrol, Guéron, Maddock, Penning, May, Pontecorvo, Seligman, Freundlich, Goldschmidt, and Placzek were unfortunately not in Montreal.

We just despatched a cable to you which explains the position concerning the Theoretical Physicists. I am afraid that we are there in very great difficulties. You know what close friends Placzek and I are. The fact that the collaboration which we started in Copenhagen might have to be interrupted for quite a while will be hard for both of us. I see many of Placzek's reasons, and I am afraid that in this case we

---

[58] Courtesy of The Public Record Office, Kew, UK. Classification AB 1/144. For a full version see M.M.R. Williams, Progress in Nuclear Energy, Vol. 36, No. 3, pp. 239-322, 2000.

will have to bear the consequences of the policy of weakness which was discontinued too late. Of course Newell and I will do whatever we can to make Placzek agree to stay and we hope that Chadwick will help us in this endeavor, but I am not very hopeful of succeeding. [...]

I discussed the above letter with Placzek who would have liked me to wait somewhat longer before dispatching it since the discussion with Chadwick might clear up some points. I told him that I did not think myself justified in leaving you uninformed of these difficulties. As a matter of fact we have tried for two weeks to see Chadwick and did not succeed. Placzek asked me to tell you that whilst the letter represents most of the facts correctly it does not mention at all that his reason for wanting to shift to other work is that he thinks that he and his team could be used in the next year or 18 months to more advantage in other fields of Tube Alloy work. He wants it to be very clearly understood that he does not want to abandon the Directorate of Tube Alloys, but he thinks it is his duty to ask to be employed where he thinks he could be most useful. I add this since he asked me, although I am convinced that you would have understood this point without my making it clear.

Yours sincerely,

H. Halban

☙

Commentary: In the subsequent letter sent the next day, March 30, 1944, von Halban continues:

> The position of the Theoretical Physicists is as follows. Marshak has resigned and would like to try to get employment in site Y [at Los Alamos] as a U.S. citizen. We know privately that Courant will get in the near future an interesting offer from another American group which we fear he will accept, particularly since were not yet able to increase his salary. Apart from these two cases, the whole Division has asked Placzek several months ago to get them as an entire group into immediate war work. Placzek asked them then to be patient. He is however of the opinion that in the case of construction in England, his group would not be needed for at least a year. On the other hand he has little confidence that anything

will come out of the present Washington discussions. He therefore informed me that he would like to discuss with Chadwick how he and his division could be shifted immediately to other work in the Tube Alloy field. In this case Marshak would stay with the group. Such a shift might enable us to get Placzek and part of his staff at a later date back into Polymer work.

I am extremely depressed about Placzek's decision although I understand him well. He knows about yesterday's visit only through cautious telephone conversations but does not think it changes his attitude. My opinion is that we need permanently a first-class theoretician with a well organized staff if any work is to be done here. In case of work in England we would need him permanently as from a year after the beginning of construction of the Polymer factory and in the meantime for consultation. Placzek thinks that Volkoff or Courant could do for the intermediate period of a year if they were willing to come. This would of course be better than nothing but in my eyes not sufficient.

I do not see much hope to hold Placzek since his misgivings concerning the speed of getting large scale work started here and straightforward collaboration organized cannot be contradicted after our past experience. [...]

✳✳✳✳✳

Gian-Carlo Wick to Rudolf Peierls [59]

Rome

September 14, 1945

[...] I am writing a book with Amaldi. I have left aside until now the chapter on the theory of resonance phenomena and so on; I expected that the paper by you, Bohr and Placzek would oblige me to change much of what I might have written. Has the paper ever appeared? Will it ever appear? Which subjects would be especially affected by it? I should be very indebted to you, if you were so kind to answer, even very briefly, to these questions. [...]

---

[59] Sabine Lee, Volume 2, page 17.

AMERICAN-SOVIET SCIENCE SOCIETY

Affiliated with

NATIONAL COUNCIL OF AMERICAN-SOVIET FRIENDSHIP

114 East 32nd Street
New York 16, N. Y.
Murray Hill 3–2082
November 14, 1945

Prof. G. Placzek
Cornell University Ithaca, NY

Dear Prof. Placzek:

We are sending you under separate cover the *Journal of Physics* which we received from the Soviet Union with the request that it be forwarded to you.

Sincerely yours,

Ignace Zlotowski
For the Executive Committee[60]

∽

January 15, 1946

American-Soviet Science Society
114 East 32nd Street
New York 16, NewYork.

Dear Sirs:

I take great pleasure in thanking you for the copy of the *Soviet Journal of Physics* which you were kind enough to send me. It made most interesting reading.

Yours sincerely,
G. Placzek

∽

---

[60]Courtesy of the AIP Niels Bohr Archive.

Placzek to Chandrasekhar [61]

Presumably, Los Alamos
Undated, circa 1946

Professor S. Chandrasekhar,
Yerkes Observatory,
Williams Bay, Wisconsin

Dear Chandrasekhar,

Many thanks for your two letters and the manuscript. I must apologize for not having acknowledged its receipt right away, but I've been traveling and have been very busy since my return. So far I have only been able to look at it rather superficially, but I have told Edward Teller about it and, before his departure for Chicago some time ago, he was interested in discussing the matter with you and I had hoped he would have seen you already by now.

I am afraid it will take me still some more time to understand the matter thoroughly enough to come to a conclusion. Whenever I start doing something about it, I am interrupted by some kind of emergency, but I very much hope to get through with it in the end.

I hope you do not need the manuscript right now, but if this should be the case, please let me know. I would then send it back to you immediately and you might return it to me later.

With best regards, sincerely yours,

G. Placzek

---

[61]Courtesy of the AIP Niels Bohr Archive.

Niels Bohr to Rudolf Peierls [62]

UNIVERSITETETS INSTITUT FOR
TEORETISK FYSIK
Blegdamsvej 17, Kobenhavn

March 7, 1946

Dear Peierls,

Thank you very much for your kind letter which recalled so many pleasant remembrances of the time we all spent together in Los Alamos where you and your wife looked after Aage and me with such great kindness. It was a great pleasure to learn from your letter that you are now back in Birmingham and, after your great contribution to the war work, are able again to concentrate on general scientific problems and cooperate with Oliphant and Moon.

I shall certainly bear in mind your offer for a time to take up in your group some young promising physicist for whom a stay with you will be such a profitable experience. We are also here trying to reorganize our experimental as well as theoretical work and I am myself endeavoring to complete the various investigations which I had to leave unfinished on my escape from Denmark.

Among these I have an especially bad conscience about our common work with Placzek and I wonder whether there might soon be an opportunity where we could meet again and look properly into it. I need not say how great a pleasure it would be to us all here if you would soon be able to visit us. My own plans are somewhat unsettled, but there is the possibility that I might come to England sometime in May, and as soon as I can survey my obligations I shall, of course, write to you again.

With kindest regards from us all to your wife and the children, and also to common friends in Birmingham,

    Yours,

Niels Bohr

∽

---

[62]Courtesy of Bodleian Library Archive, Oxford University.

Placzek to Marshak[63]

March 12, 1946

Professor R. E. Marshak,
Rochester, New York

Dear Bob,

I have just returned from a trip East and am finding your letter here.

Sorry to have missed you this time; I didn't realize you were back in Rochester already. Anyway, there is no harm done because I was rather in a hurry this time, spent most of the time in Chicago and Schenectady and only two days in New York. I expect to be East again towards the end of April or the beginning of May and I hope we can arrange to meet at that time, unless you prefer to come out here before then.

I have forwarded your memorandum to Muncy. In regard to the reports, A-25 has been sent out to you, our copy of A-4 was off the site in connection with its impending publication. (This is a report which was done before the existence of any official government project and is, therefore, not subject to declassification.) It will come out in the Physical Review. I have it now back here and will see to it that it reaches you soon.

In regard to declassification of project reports, we are doing what we can to expedite matters, but things are still in the initial stages and seem to move rather slowly. You will hear from us as soon as something is decided.

I do not know any more than Viki about things here since I left a few days after he did. In general, I tend to agree with his outlook.

I am writing this in a hurry so Gerhard can take this letter out to you.

Sincerely,

G. Placzek

∞

---

[63] Courtesy of the AIP Niels Bohr Library.

Placzek to Bohr [64]

P.O. BOX 1663
SANTA FE, NEW MEXICO
April 23, 1946

Professor Niels Bohr,
Institut for teoretisk Fysik,
Blegdamsvej 15,
Copenhagen, Denmark

Dear Professor Bohr,

The Director of the Los Alamos Laboratory, Dr. Bradbury, has asked me to write to you about a conference on nuclear physics which is to be held at Los Alamos this summer. The participants in the conference will be physicists who have worked in the various branches of the project during the war, as-well as physicists who have no connections with the project.

We should all be very happy if you would find it possible to attend this conference. It has been provisionally scheduled for the middle of August, but if you were at all interested, we would be most anxious to fix the date in such a way as would make it most convenient for you to attend. I should be most grateful if you could let me know about this matter.

I take it that you are very busy now in getting normal work at the Institute under way again. We have all been very interested in reading the reprints which we have recently received from Copenhagen, on the work of the Institute during the war.

Los Alamos has changed very much during recent months since most of the old people have left, especially in the Theoretical Division, I intend to finish my work here in June and am going to join then the research laboratory of General Electric at Schenectady. You have probably heard about the recent discovery of mesons produced by the gamma rays from their 100 MeV betatron. There seems to be now general agreement that the tracks observed are genuine meson tracks. Many statements have also been made about the masses of these particles but they seem to be still somewhat controversial. So far, cloud chamber pictures have only been made with a magnetic field of insuf-

---

[64]Credit: Niels Bohr Archive, Copenhagen.

ficient strength and the observed curvatures might possibly be falsified by multiple scattering. There was quite some discussion on this subject which seems to me rather futile since it would be very easy to decide this question by experiments with an adequate field. In fact, such work is now in progress and accurate mass spectrographs for mesons are also under construction. The work has been somewhat delayed by a strike which interrupted access to the Laboratory for almost three months at the beginning of the year but things are now in full swing again and I think that better results can be expected soon. I hope to keep you informed about them. You have probably also heard about the new experiments by Bloch and Purcell on nuclear magnetic moments and relaxation times. Their methods seem to be extremely simple and elegant and, due to the fact that the relaxation times seem to be very much shorter than was believed, they work exceedingly well.

I am very much looking forward to hear from you. With best regards and all good wishes to yourself, your family and all the friends,

Yours sincerely,

G. Placzek

∞

Bohr to Placzek[65]

Gl.Carlsbergs, Valby-Copenhagen
April 30, 1946

Dear Placzek,

I thank you very much for your kind letter with, the invitation from Dr. Bradbury to attend the conference on nuclear physics in Los Alamos this summer. It should certainly be a great pleasure for me to come and meet you and other friends again in the surroundings where Jim and I had so unforget[table] an experience, but I am sorry that it will hardly be possible. The situation is that I have already accepted invitations to attend the meeting of the American Physical Society in Now York on September 19-21 and the Bicentennial meeting in Princeton immediately thereafter. Due, however, to my obligations here with the reorganization of the work after my long absence, it will be difficult for

---

[65]Credit: Niels Bohr Archive, Copenhagen.

me this time to make too long a stay in the U.S.A., but still I hope very much that we shall have opportunity to meet and especially to discuss plans for future cooperation. I hope also that my wife will be able to accompany me and meet our many old friends again, and she looks forward not least to seeing you and your wife.

With kindest regards and best wishes from us all,

Yours,

Niels Bohr

∽

G. Placzek to Rudolf Peierls [66]

Schenectady, March 7, 1947

Dear Peierls,

Many thanks for your detailed letter and the statement of the Council of the British Atomic Scientists, which I have read with great interest. It tends to show that, if even scientists of one nation cannot agree in the fundamentals of the matter we need not be surprised to see the politicians land in the present hopeless mess.

You are probably aware that the statement has been extensively misquoted by Mr. Gromyko in yesterday's session of the Security Council.[67] This is of course not your fault since even if it had been possible to write it from a more unified point of view, this would not have afforded protection against distortions. I hope you can get a verbatim copy of the speech, it was rather long and the papers left out the most interesting passages. I happened to listen to it on the radio and it was rather characteristic to hear Gromyko state in terms which were clearly sincere conviction that the idea of an inspector who could freely travel in a "foreign" country and even fly over it, could not have been meant seriously.

The Russians, of course, have by no means a monopoly on such a mentality; the recent statement by Senator Taft, warning against the

---

[66]Sabine Lee, Volume 2, page 74.

[67]At that time Andrei Gromyko (1909-1989) was the Soviet Permanent Representative at the UN Security Council. On 5 March 1947 Gromyko gave a speech in the Security Council denouncing American nuclear policy. See *New York Times* of March 5, 1947, p. 16. From 1957 to 1985 Gromyko was the USSR Foreign Affairs Minister.

danger that communists might "infiltrate" into an international control agency, is on a very similar level.

Enclosed [is] a pictorial record of yesterday's meeting, taken from today's *New York Times*. You will probably enjoy the subscript.

I am afraid I cannot entirely share the Olympian detachment of your American military friend who hopes that chances of an agreement might be better once stories of successful Russian bomb manufacture begin to circulate. I am rather inclined to believe the opposite. But unfortunately, we shall see.

Clayton[68] (from Chalk River) asked me for a copy of the table of the exponential integral for complex argument, so that he could forward it to you. Unfortunately I have no copy here. I believe I left one at Montreal, but whether or not this is so, they cannot find it. The rest of the copies is in Carlson's hands[69] at Los Alamos and to extricate one from there, under present circumstances, might take longer than to have the whole tables recalculated. I therefore recommended to Clayton to get in touch with Lowan, (Math. Tables Project) who might have a spare copy or could perhaps get one reproduced.

I hope you will keep me informed on the result of your and Skyrme's investigations re Wigner Dispersion formula, and gauge invariance of Weisskopf's logarithmic divergence.

Newspaper reports here inspired, in some credulous souls, the hope that Schrödinger might have discovered the laws of the universe. Mr. W.L. Lawrence was speeded into action. Jumping at the opportunity to act as a great patron saint of science, he got hold of Schrödinger's manuscript, had it photostated at the *Times* expense, and distributed far and wide throughout the country. Of course, it turned out to be bunk. A lesson for the above-mentioned souls showing that the more traditional channels of scientific information still seem to be fully adequate.

With best regards to you and Genia,

Yours sincerely, G. Placzek

∽

---

[68] Henry H. Clayton (1906-89), English-born physicist who had studied at the University of British Columbia. In 1945 he joined the Montreal Laboratory and moved to Chalk River in 1946. In 1950 he became head of the theoretical physic branch, a position held until his retirement in 1969. - SL

[69] Bengt Carlson (1915-2007), Swedish-born physicist who had studied at Stockholm and Yale. He worked in Placzek's group at Montreal from April 1943. -SL

Rudolf Peierls to George Placzek [70]

[Birmingham]
April 24, 1947

Dear Placzek,

Thank you for your letter of 7th March and the highly amusing picture of Uncle James.[71] Meanwhile we have had the full text of Gromyko's speech and, as you say, it is a very depressing document even if it has its amusing points. If, however, one looks at the earlier statements made at one time or another by Gromyko and Molotov and the many contradictions in them, it is clear that the fact of their saying decidedly "no" today rules out as little the possibility of their saying "yes" tomorrow as vice versa. As far as the British Atomic Scientists' Association is concerned, we feel we ought to appoint Gromyko honorary publicity manager or something for the way he has put us on the map. With one exception his quotations are quite correct, although, of course, torn from the context they give the wrong impression.

I saw the articles in the New York papers that followed Gromyko's speech by a few days in which other passages from our memorandum were quoted, in this case, of course, carefully picking those parts in which we agree with the American point of view.

I do not quite share your view that the document was quite futile; it is, of course, a question whether the present situation can be remedied at all by a rational discussion in the Atomic Energy Commission of the Security Council and it is likely that if any progress is made it may depend on discussions behind the scenes and on factors that have nothing to do with the problem in hand, but to the extent that it was worthwhile to keep rational discussion going I do believe that we have raised points where obstacles to the acceptance of the details could be removed.

Even more amusing, perhaps, are the Senate hearings about Lilienthal.[72] I enjoyed particularly the picture of Groves releasing the Smyth report against his better judgement because of the bullying of scientists. On the question of tables of the complex potential integral, it

---

[70]Sabine Lee, Volume 2, page 79.

[71]See Placzek's letter of March 7, 1947, on page 184.

[72]On 4 February 1947 David E. Lilienthal had issued a controversial statement "This I Deeply Believe," also known as the "Credo." See David E. Lilienthal Papers, Volume 1, 1900-1949, Box 118-119. Princeton University Library. -SL

will be interesting to see what ultimately comes out of the revolutions in the wheels of the machine. In practice, as you may remember, there was a blueprint of the original typescript of those tables here and that is still accessible to us. We can therefore await further developments with patience [...]. On the general dispersion formula we have not yet made much progress beyond one small point:

In the Kapur-Peierls method one expands the whole wave function inside the nucleus in a series of complex eigenfunctions. As a result, one gets a potential scattering term representing the scattering from a hard sphere and alongside with it a dispersion sum which seems to converge rather slowly. Skyrme has pointed out that one could equally use an alternative procedure, namely to expand not the whole of the function, but its difference from the incident plane wave. In that case the potential scattering term represents the Born approximation and, in particular, the artificially chosen nuclear radius no longer appears explicitly. For high energies of the incident neutron it seems fairly clear that in this formula the contribution from the distant terms of the dispersion sum are smaller than in the alternative case. This will certainly be proved when the energies are so high that the Born approximation becomes a useful approximation and is likely to be true even for somewhat lower energies.

Once one has recognized that there is this amount of freedom in the method, a third alternative immediately suggests itself and that is to start by solving a one-body problem with a complex potential representing, in other words, the nucleus as a black rather than a reflecting sphere, and after subtracting the wave functions ascertained from the whole wave function expand the residue in a series as before. This means using the theory of a black nucleus as developed by Bethe as a starting point. Evidently this will lead to a formula in which the potential scattering appears explicitly as that due to a black sphere and added to it there will be again a dispersion sum.

On physical grounds I suspect that this description would be the most convenient one, i.e. it would make the dispersion sum converge more rapidly than the alternatives. Of the three possibilities this third one certainly is the most complicated one mathematically and there seems to be no obvious way either to prove that it is a good approximation or to decide the value of the absorption coefficient which ought to be used. I would much appreciate your comments on this situation. We have not been able to get much thrill out of the recent papers by

Wigner or by Weisskopf on this subject.[73]

I agree with your comments on the wisdom or otherwise of letting newspapers distribute scientific manuscripts, but not that the traditional channels are very adequate at the moment. Printing is deplorably slow, particularly in this country, and one has to rely largely on casual gossip and, while this is apt to make more sense than the efforts of Mr. W. L. Lawrence, it is hardly more efficient.

With best wishes,

[R.E. Peierls]

P.S. I am working on a plan to have a small theoretical conference at Birmingham this summer, to deal with the fundamental difficulties and with elementary particles. The likely dates are July 23rd to 26th, and I hope to get some of the theoreticians in this country and a fair number from the Continent to take part. Our funds do not make it possible to pay transatlantic fares but if, by any chance, you were intending to be on this side this summer, we would look after your board and lodgings, probably in the University hostel here, during the conference. You will get a more official notice in due course but I thought I would let you know at once that this is likely to happen. I need not say that we should be delighted to see you here.

∽

---

[73]E.P. Wigner, "Resonance Reactions and Anomalous Scattering," Phys. Rev. 70, 15-33 (1946); "Resonance Reactions," Phys. Rev. 70, 606-618 (1946); E.P. Wigner and L. Eisenbud, "High Angular Momenta and Long Range Interaction in Resonance Reactions," Phys. Rev. 72, 29-41 (1947); H. Feshbach, D.C. Peaslee and V.F. Weisskopf, "On the Scattering and Absorption of Particles by Atomic Nuclei," Phys. Rev. 71, 145-58 (1947).

Figure 3.3   George Placzek, Genia Peierls, Klaus Fuchs and Rudolf Peierls. Tentatively, late 1940s, London. From Sabine Lee, *Sir Rudolf Peierls: Selected Private And Scientific Correspondence*, (World Scientific, Singapore, 2007), Volume 1, page 434b.

Rudolf Peierls to Niels Bohr [74]

Birmingham,
February 6, 1948

Dear Uncle Nick,[75]

I should probably have written before to say that I had a brief opportunity while I was in America to discuss the draft of our paper [76] with Placzek. Placzek raised a number of small points that might need amending, but it seemed to us that all these could be taken care of by alterations of a few words, they could well wait until we knew your reaction to the main outline. It has occurred to me that one of these points might be causing you difficulty and that it might save you trouble to draw your attention to it. It concerns the derivation of the Breit-Wigner formula.

The derivation which I have sketched is valid only for the part of the resonance curve for which the kinetic energy of the emerging neutron (or other particle) differs only by a small fraction from its value at resonance. It does not cover either cases in which the width of the resonance level is comparable to the kinetic energy of the neutron at resonance or the cross section for thermal neutrons ($1/v$ law). We tried to see whether it was easy to generalize the derivation so as to cover these cases as well, but we felt that this was not possible without spoiling the transparency of the argument, but that it was preferable, therefore, to leave the derivation as it stands and merely to make clear to which category of problem it is applicable. Since the purpose of the paper is mainly to deal with high energies, it would be quite reasonable to use an argument which is not appropriate for very low velocities.

I am afraid that local arrangements made it necessary to make a decision on our plans for the summer about which I wrote to you before and we have decided to go ahead with a conference here in the week starting 20th September. This will be a joint affair of the Physics and Mathematical Physics Departments. I very much hope that in doing so, we are not clashing too badly with any plans you have in mind.

---

[74]Sabine Lee, Volume II, page 124.

[75]Nickolas Baker was the code name of Niels Bohr at Los Alamos.

[76]Rudolf Peierls had been to the US for a conference, and on that occasion met George Placzek with whom he discussed the Bohr-Peierls-Placzek "long paper." Peierls appears to have left a copy of the draft with Placzek who commented on it and passed the revised manuscript on to Bohr, when the latter came to Princeton in May 1948. See R. Peierls, "Introduction," in Niels Bohr, *Collected Works*, Vol. 9, p. 51. -SL

With very best wishes to all friends in Copenhagen,
Yours sincerely,

R. E. Peierls

∞

Niels Bohr to Rudolf Peierls [77]

UNIVERSITETETS INSTITUT FOR
TEORETISK FYSIK
Blegdamsvej 17, Kobenhavn
August 22, 1949

Dear Peierls,

It has been a pleasure to me that it is now arranged that Lindhard will be with you next year. I am sure that it will be a great experience to him and I also hope that it will mean a still closer cooperation between our groups. As a small beginning I reckon that Lindhard's stay with you will be helpful in completing our old work with Placzek.

In the last weeks I have gone through the old manuscripts with Lindhard and discussed with him the latest progress as regards nuclear constitution and in particular the success of the method j of considering the binding of the nucleons separately in the nuclear field. I realize that one sometimes takes the drop model too literally and, to clear my thoughts, I have written down a few tentative comments of which I shall be very glad to hear your opinion. They do not contain much new, but I feel that the development gives a simpler basis for the treatment of the problems of nuclear reactions and removes doubts as regards the conclusions to be drawn from dispersion theory and detailed balancing. As soon as I have time, I will try to incorporate such views in our old manuscript and will, if not before, give it to Lindhard when he leaves. This summer I have been busy with the preparation of a series of lectures on general topics, which I shall deliver in Edinburgh in the autumn and have also worked with Rosenfeld on the completion of our work on the measurability of field and charge quantities. It has come out that the situation is just as required by Schwinger's formalism and that it is simpler than assumed by Heisenberg in the respect that charge

---

[77]Sabine Lee, Volume II, page 103.

fluctuations are well defined in sharply limited space-time extensions, just like field fluctuations. Also this work I hope to complete in the autumn months. As you may understand, it will be quite a busy time for me and, if it is not too inconvenient for you, I should be glad if Lindhard could stay here until I leave for Edinburgh in the middle of October or in any case till the end of September.

With kindest regards and best wishes to your family and yourself from us all,

Yours Niels Bohr

∽

Niels Bohr to Rudolf Peierls [78]

UNIVERSITETETS INSTITUT FOR
TEORETISK FYSIK
Blegdamsvej 17, Kobenhavn
May 25, 1950

Professor R. Peierls,
Department of Mathematical Physics,
The University, Edgbaston, Birmingham 15

Dear Peierls,

I thank you for your kind letter of May 16th and hasten to answer that Dr. Barker shall be most welcome indeed to work with our group for the next academic year. I was very interested in what you wrote about his abilities and about his work. We are also here just now occupied with the problems of nuclear reactions on such lines and expect in a few weeks a visit of Dr. Hill from the U.S.A. who has worked with Wheeler and I on the fission problems, and I hope with him to soon complete a paper just on the relationship between the drop model and an individual particle model in such respects.

In Paris you probably have heard that considerable progress has recently been made in the understanding of the relationship between nuclear shape and nucleon binding. This gives not only a far-reaching quantitative account of the quadrupole moments of nuclei, but implies also a coupling between the excitation of individual nucleons and the

---

[78]Sabine Lee, Volume II, p. 230.

oscillations of the whole nucleus which offers a general understanding of the properties of the formation of the compound state in nuclear reactions. Our old work with Placzek has also been much on my mind, but both Placzek and I were so occupied in Princeton with other work and interests that we only had a few discussions about this paper, which I hope we all can complete as soon as I have got the work with Wheeler and Hill off my hands.

It has been a great pleasure to learn that Lindhard has had such a good time in Manchester, and I look forward to see him soon again and go on with the work on the charge of fission fragments, about which Lassen's experiments have given such interesting results. Would you kindly greet Lindhard for me and say that I shall be glad to learn about his plans. I have not written myself to him because I found so very much to do in the first weeks after my return from Princeton.

I cannot close this letter without remembering the very noble and moving letter you sent me at Princeton about the tragic case which has brought so much anxiety into wide circles. We are certainly not living in a pleasant world, but in spite of it all I keep up the hope that we shall see better times before it is too late.

With the kindest regards from my wife and I to your whole family and yourself,

Yours ever,

Niels Bohr

∽

Freeman Dyson to Rudolf Peierls [79]

Princeton
June 24, 1950

[...] Placzek said the reports of his death had been partly exaggerated. He is, in fact, walking around and working, and seems to be in as good a state as last year. They are staying here through the summer and living quietly because they are afraid his kidney may give more trouble, but at present it is behaving well. Placzek has been working

---

[79]Sabine Lee, Volume II, p. 239

all this year with a young Dutchman called Nijboer[80] on the details of the interference phenomena in neutron scattering, even a lot of numerical work has been done. He and Els send you and the family all their best wishes. [...]

∽

UNITED STATES
ATOMIC ENERGY COMISSION
Washington 25, D.C.

P. O. Box 1663
Los Alamos, N.M.
November 30, 1950

Dr. J. G. Beckerley U. 8.
Atomic Energy Comission
1901 Constitution Avenue
Washington 25, D. C.

Dear Jim:

This is to inform you that Both Drs. Bethe and Placzek have reviewed LADC-887 and have no objection to its declassification under Topic 1-100 of the December 1950 Guide.

I understand that the Los Alamos Scientific Laboratory is submitting this article for publication in the Review of Scientific Instruments.

The Oak Ridge Office will receive reviewer forms as soon as new ones have been printed in accordance with the Issuance of the new Guide.

Yours,
Frederic de Roffmann[81]
Security Committee on Neutron Diffusion and Reactor Theory

cc: H. A. Bethe
G. Placzek, R. C. Smith

---

[80]The result of their work was published as G. Placzek, B.R.A. Nijboer and L.V. Hove, "Effect of Short Wavelength Interference on Neutron Scattering by Dense Systems of Heavy Nuclei," Phys. Rev. 82, 392-403 (1951). -SL

[81]Credit: Shelby White and Leon Levy Archives Center, Institute for Advanced Study, Princeton, NJ, USA.

Bohr to Placzek[82]

UNIVERSITETETS INSTITUT
FOR TEORETISK FYSIK
COPENHAGEN, DENMARK
TELEGRAMS: PHYSICUM

December 7, 1950

Professor G. Placzek,
Institute for Advanced Study,
Princeton, New Jersey

Dear Placzek,

In the days of July 6-10, 1951, it is planned under the joint auspices of the International Union of Pure and Applied Physics and the Institute and under the sponsorship of UNESCO to arrange a Conference on problems of quantum physics in Copenhagen.

The restricted facilities and the endeavors to have proper informal discussions like in the earlier Copenhagen conferences will only permit to invite a limited number of physicists to attend the Conference, but we especially want to extend a cordial invitation to the old friends and collaborators of the Institute.

During the Conference you will be a guest of the Institute which will arrange for housing accommodation. Unfortunately, however, we cannot cover traveling expenses for visitors from countries overseas, but we hope that the participants from these countries will be able to obtain grants for their traveling expenses from their own institutions or from other public sources.

For the sake of our arrangements we should be glad, at your earliest convenience to learn whether you will be able to accept our invitation provided practical circumstances will permit your journey.

Hoping that we shall have the pleasure of seeing you at the Conference, and with kind regards and best wishes,

Yours,
Niels Bohr

---

[82]Credit: Shelby White and Leon Levy Archives Center, Institute for Advanced Study, Princeton, NJ, USA.

SECRET TRANSMITTAL
United States
ATOMIC ENERGY COMISSION
Washington 25, D.C.

May 15, 1951[83]

Dr. G. Placzek
Institute for Advanced Study
Princeton, New Jersey

Dear Dr. Placzek:

I have just received from the Chalk River group a copy of TPI.45. It is my understanding that you are interested in having this report declassified.

The Canadians have pointed out that pages 24 and 25 of the report refer to CP-644 and the fast fission correction in lattice calculations. Their specific question is whether we agree to declassifying the report with this material included. I hesitate to bother you with this but, since I believe you are familiar with the report and have some interest in it, perhaps you would comment on whether we should declassify the report (including pages 24 and 25) and go along with the Canadian request. As the fast fission correction was discussed at the recent New York meeting you are also in a position to recognize whether its release would be consistent with the consensus of the meeting.

Kindly return the report to us at your convenience (as it belongs to the Canadians).

The draft proposals to change Section 10 of the Atomic Energy Act, which I sent you a few weeks back, are having a difficult time. The Department of Defense has some reservations on the present version so I presume it will be rewritten.

Best personal regards,

James G. Beckerley
Director of Classification

⚮

---

[83]Credit: Shelby White and Leon Levy Archives Center, Institute for Advanced Study, Princeton, NJ, USA.

UNIVERSITA DEGLI STUDI – ROMA
ISTITUTO DI FISICA
"GUGLIELMO MARCONI"

May 26, 1951 [84]

Dr. G. Placzek
Institute for Advanced Study
Princeton, N.J.

Dear Friend:

On Sept. 26, 1951, E. Fermi has his 50th Birthday.

In order to commemorate this anniversary we have thought to invite his direct pupils and scientists that have worked with him or in his Institute to subscribe to a fund directed to the institution of a "Fermi medal." We would like to have the medal given every year by an authoritative body such as the Executive of the International Union of Physics or their delegates (among which we shall try to have Fermi himself) to a young (under 40) physicist of great promise.

We hope that the medal could be of some help to further the career of the recipient.

We have sent this circular to the enclosed list of people hoping that we have not made too many omissions. If you know of anybody else who should be contacted please write him directly.

The funds collected will be used to coin as many gold medals as possible (about 50 gr. per medal) and the medals will be kept by the *Accademia dei Lincei* in Roma, the name of the recipient will be engraved every time on the medal.

We would like to accumulate not less than $1000 in order to have about 12 medals. This means a subscription of $20. If you wish to contribute please mail checks to Prof. S.K. Allison, Univ. of Chicago, Chicago Ill. for U.S.A. subscribers. European subscribers may send their contributions to Prof. E. Amaldi, Istituto di Fisica dell'Universita Roma.

Very sincerely yours,

E. Amaldi
E. Segré

---

[84]Credit: Shelby White and Leon Levy Archives Center, Institute for Advanced Study, Princeton, NJ, USA.

LOS ALAMOS SCIENTIFIC LABORATORY
Classified Materials Receipt Form

69460

To: ___George Placzek___

Date___July 20, 1951___
(and time of dispatch)

RECEIPT OF CLASSIFIED MATERIALS

ORIGINAL—To be signed personally by the recipient or his authorized delegate and returned to sender.
DUPLICATE—To be retained by recipient.
TRIPLICATE—To be retained by sender for suspense file and destroyed when signed original is returned.

I have personally received from (sender)___Mail &Records___
P. O. Box 1663, Los Alamos, New Mexico, the material as identified below. I assume full responsibility for the safe handling storage, and transmittal elsewhere of this material in accordance with existing regulations.

The material, including enclosures and attachments, is identified as follows: (In identifying material, avoid any reference which might cause the receipt form to become CLASSIFIED).

Description	Ref. or File No.	Date of Document	From	Addressed to
Copy 53A of Secret Minutes	SRR-282	7/14/51		

SO: Signed Original; CC: Carbon Copy; PC: Photostatic Copy; TC: Typed Copy

T-606
(Postal Registry Number)

(Signature of Recipient)

7/23/51
(Date Received)

Form 262—25M—4-50

RECIPIENT'S COPY

George Placzek to G. von Dardel[85]

December 20, 1951

Dr. G. von Dardel
Aktiebolaget Atomenergi
Department of Physics
Drottning Kristinas Vag 47
Stockholm Sweden

Dear Dr. von Dardel:

May I acknowledge receipt of your letter of December 13. I am enclosing herewith a paper on neutron scattering <u>heavy</u> nuclei which will be published in the Physical Review.

The corresponding theory for light nuclei and in particular for protons is somewhat less complete. With certain qualifications which are not very important for practical puposes it can, however, be shown that also in in the case of protons the collision may in good approximation be treated as if the nucleus were free, provided the neutron energy is large compared to the level separation. The asymptotic expression for the cross section

$$\sigma = \sigma_{\text{free}} \left\{ 1 + \frac{1}{3} \frac{K_{\text{Av}}}{\mu E_0} \right\} \tag{1}$$

(notation as in enclosed paper) will therefore also hold in the case of protons. While for heavy nuclei the above expression (apart from interference terms unimportant in the case of protons) is asymptotically exact, it represents in the case of protons only an average around which the cross section may in certain cases exhibit undamped oscillations, of the same size as the second term in (1). The example of an oscillator where this is the case has been discussed in Section 6 of the enclosed paper. To supplement this discussion, I am also enclosing, on a separate sheet, data showing the variation of the quantity $c(n_0)$ defined by equation (6.10) for the case $\mu = 1$. In the case of molecules these oscillations are quickly damped by the rotation as has been confirmed by an example discussed in Messiah's paper (Phys. Rev. 84, 204, 1951).

The average kinetic energy K_{Av} which consists of translational, rotational, and vibrational parts, the latter including the zero-point energy, can be calculated from spectroscopic data.

---

[85]Courtesy of the Shelby White and Leon Levy Archives Center, Institute for Advanced Study, Princeton, NJ, USA.

I intend to write, in the near future, another paper on neutron scattering with special reference to light nuclei, and I hope to send you a provisional copy within a few months' time. I may also perhaps come to Europe this summer in which case I should be very happy to discuss this matter with you.

Sincerely yours,

G. Placzek

∞

Bohr to Placzek[86]

UNIVERSITETETS INSTITUT
FOR TEORETISK FYSIK
COPENHAGEN, DENMARK
TELEGRAMS: PHYSICUM

April 28,1952

Professor G. Placzek,
Institute for Advanced Study,
Princeton, New Jersey

Dear Placzek,

You will have heard from Pais about the planned European cooperation in Nuclear Physics which we all hope will bring very fruitful results. Of course, the cooperation of American colleagues will be very important for the endeavors and here in the Institute we hope not least to benefit from the advice of old friends like yourself. You will also have heard about the Conference which is to take place here during the first three weeks of June to initiate the work of the Study Group to be assembled at the Institute for the coming year. I am writing especially to say how happy we would all be if it fits in with your plans to come to Europe this summer and take part in the Conference.

---

[86]Courtesy of the Shelby White and Leon Levy Archives Center, Institute for Advanced Study, Princeton, NJ, USA.

Margrethe joins me in sending you and your wife our kindest regards and best wishes,

Yours,
Niels Bohr

P.S. I need not say how grieved we were over the death of Kramers, which means so great a loss to physics and to his many friends. Here, where he came as a young student and was a cornerstone in our group for so many years, it is difficult to reconcile ourselves with the idea that we shall not see him anymore.

∽

George Placzek to G. van Wijk [87]

May 19, 1952

Professor W. R. van Wijk
Laboratorium for Natuuren Weerkunde
Duivendaal 2
Landbouwhogeschool
Wageningen, Holland

Dear van Wijk,

Many thanks for your letter of April 30th. Kramers' death has been a great shock. I am particularly grateful to you for the report about the funeral. De Boer seems to have expressed what so many of us felt.

I have thought about the question of DeWitt's visit to America, and I have also discussed it with Pais. While you are certainly right in thinking it preferable for DeWitt to spend the time in America in contact with physicists rather than in a purely agricultural place, it seems to us that the Institute would hardly be the place most suitable for his purposes. The physicists here work on nuclear physics, meson theory, statistical mechanics and related problems in a rather abstract style and consequently DeWitt would feel quite isolated and would not find the contacts so necessary for his work even if I were to take some interest in it. It seems to me therefore that DeWitt should go to a

---

[87]Courtesy of the Shelby White and Leon Levy Archives Center, Institute for Advanced Study, Princeton, NJ, USA.

place where there is same emphasis on and interest for applied physics. I believe that there are many such places in the United States, and I have tried to inquire which of these would be most suitable for DeWitt. I feel however that the suggestions I received (among the places mentioned were the University of Wisconsin and M.I.T. in Cambridge) are not precise enough to enable me to make a definite recommendation to you. Under these circumstances it seems to me that it would be by far the best for you to write to Hendricks after all. Kramers has already talked to him about your work and thus he will be able not only to advise you competently but perhaps also to help in finding a way to supplement DeWitt's travel stipend.

I have tried to answer your letter as quickly as possible – it is for this reason that I am writing in English – and I hope there was not too much delay.

With best regards,

Yours sincerely, G. Placzek

George Placzek to Niels Bohr [88]

June 5, 1952

Professor Niels Bohr Universitetets
Institut for Teoretisk Fÿsik
Blegdamsvej 15
Copenhagen, Denmark

Dear Bohr,

May I thank you for your letter of April 28th and for your kind invitation to attend the Conference.

I am planning to come to Europe this summer but I was not able to arrange for leaving early enough to take part in the Conference. I should very much like, however, to visit Copenhagen at a later date if convenient for you, in order to see you and to keep in touch with the Study Group. I understand from Weisskopf that he plans to come to Copenhagen around September 3rd, and I thus should like to inquire if it would be all right with you if I came at about the same time. M? addresses in Europe are not quite fixed as yet but from the end of June on mail will reach me c/o Dr. R. Planiol, 56 rue d'Assas, Paris 6^e.

I have just learned from Robert [Oppenheimer], with regret, that you will not be able to spend the next term with us in Princeton, and this is of course one more reason for me to come to Copenhagen. With kindest regards and all good wishes to you and your family, also from my wife,

Yours sincerely,
G. Placzek

P.S. Kramers' memory is very much with us here where he spent the last few months of his life.

❧

---

[88]Courtesy of the Shelby White and Leon Levy Archives Center, Institute for Advanced Study, Princeton, NJ, USA

Rosbaud to Placzek[89]

PERGAMON PRESS LTD.,
242 MARYLEBONE ROAD,
LONDON, N.W.1
Ambassador 3421.
September 8, 1953

Dr. G. Placzek,
918 St. Davids Lane, Schenectady,
N.Y., U.S.A.

Dear Dr. Placzek,

I have not heard from you for years and I would very much like to see you again. I am arriving in New York on the Ile de France on October 13 and I shall be in Schenectady probably on October 22nd. I have written to Hollomon whom I would like to see, but I have no idea where you are. The address I have is from 1947. Should this letter reach you would you please drop me a line telling me where I can find you.

My work is developing very well, and I am sure you may have seen some of our journals and books without knowing perhaps that they are my babies.

With best regards,

Yours sincerely,
P. Rosbaud[90]

∽

---

[89]Courtesy of the Shelby White and Leon Levy Archives Center, Institute for Advanced Study, Princeton, NJ, USA.

[90]Paul Rosbaud was a scientist who worked for the Springer Publishing House and later for the Pergamon Press and who knew all distinguished physicists of his time. A book by the American writer Arthur Kramish describes Rosbaud's anti-Fascist activity, see A. Kramish, *The Griffin*, (Boston, Houghton Mifflin Co, 1986).

BY AIR MAIL
PAR AVION
AIR LETTER
AEROGRAMME
6ᴰ 2 JUNE 1953 6ᴰ
R
1953
CORONATION
Dr. G. Placzek,
918 St. Davids Lane,
Schenectady,
N.Y.,
U.S.A.
5 Hartwell Lane
Princeton
N.J.
Second fold here
Sender's name and address: Pergamon Press,
Marylebone Road,
N.W.1.
IF ANYTHING IS ENCLOSED THIS LETTER
MAY BE SENT BY ORDINARY MAIL

George Placzek to Paul Rosbaud[91]

September 30, 1953

Dr. P. Rosbaud
Pergamon Press Limited
242, Marylebane Road
London N.W.1, England

Dear Dr. Rosbaud,

It was good to hear from you again after so many years, and I am most happy about the opportunity of seeing you in this country. Your letter addressed to Schenectady reached me with some delay; since 1948 I have been at The Institute for Advanced Study at Princeton, and I very much hope that you will visit me here soon.

Princeton can be reached from New York by an hour's train trip. Accordingly I would suggest that you get in touch with me after your arrival in New York so that we may arrange for a mutually convenient date.

Looking forward to seeing you soon, with best wishes for a pleasant crossing,

Yours sincerely,
G. Placzek

---

[91] Courtesy of the Shelby White and Leon Levy Archives Center, Institute for Advanced Study, Princeton, NJ, USA.

ATOMIC ENERGY OF CANADA LIMITED

CHALK RIVER PROJECT
CHALK RIVER, ONTARIO
CABLE ADDRESS "MOTA"

January 28, 1954 [92]

Dr. G. Placzek
Institute for Advanced Study
Princeton, N.J., U.S.A.

Dear Dr. Placzek:

Thank you for the preprints of your articles on neutron scattering by crystals. The work here has been interrupted but we are preparing apparatus to measure the neutron spectrum resulting from the scattering of cold neutrons by single crystals. In this connection the paper by yourself and Dr. Van Hove is of particular interest because of the emphasis on such experiments. The results in this paper will be very useful as a guide to the most suitable experimental arrangements. It has arrived at a most opportune time.

Yours sincerely,
D.G. Hurst

ℴ

---

[92]Courtesy of the Shelby White and Leon Levy Archives Center, Institute for Advanced Study, Princeton, NJ, USA

George Placzek to Rudolf Peierls [93]

May 18, 1954

Professor R. E. Peierls
Department of Mathematical Physics
University of Birmingham
Birmingham 15, England

Dear Peierls:

Many thanks for your letter of [May] 4th and the preview of your program. I am very much looking forward to seeing you and I will be delighted to come to Birmingham.

If it were convenient for you, I could, for example, come on June 18 or 19 and stay until about the 22nd. I would probably arrive in England on June 7th and I tentatively plan to spend thereupon some time in Oxford examining together with Squires some of his recent results (incidentally, Squires will probably come to the Institute for the next term) but this could probably be shifted or interrupted if necessary. If you do not wish to make definite arrangements now, however, we could also settle the date of my visit to Birmingham after my arrival in England. With kindest regards,

Yours sincerely,
G. Placzek

&c;

---

[93]Courtesy of the Shelby White and Leon Levy Archives Center, Institute for Advanced Study, Princeton, NJ, USA.

UNIVERSITY OF CALIFORNIA
LOS ALAMOS SCIENTIFIC LABORATORY
(Contract W-7405-Eng-36)
OFFICE OF THE BUSINESS MANAGER
P. O. Box 1663
Los Alamos, New Mexico

Dr. George Placzek
Institute for Advanced Study
Princeton, New Jersey

June 14, 1954[94]

Dear Dr. Placzek:

The Director of this Laboratory has requested that I contact you relative to the renewal of your present consultant agreement to cover the period July 1, 1954 through June 30, 1955. We are in receipt of a letter from Dr. Robert Oppenheimer, Director of the Institute for Advanced Study, authorizing us to contact you in this regard.

If this proposed renewal meets with your approval, will you kindly execute the attached Consultant Agreement #204, Supplement #6, indicating the date of your acceptance thereof and your business address. The original should then be returned to this office and the second copy retained by you. Very truly yours,

A. E. Dyhre
Business Manager

cc: Dr. N. E. Bradbury

∽

---

[94]Courtesy of the Shelby White and Leon Levy Archives Center, Institute for Advanced Study, Princeton, NJ, USA.

THE INSTITUTE FOR ADVANCED STUDY
PRINCETON, NEW JERSEY

SCHOOL OF MATHEMATICS
June 17, 1954[95]

Dr. George Placzek
Clarendon Physical Laboratory
Oxford, England

Dear George:

At a meeting of the 14 members present in Princeton today 9 were in favor of issuing a collective statement in the event of an adverse decision by the A.E.C. concerning Oppenheimer's security clearance. Two were opposed to any collective statement (von Neumann and Goldstine) and three abstained from expressing a final opinion (Morse, Montgomery, Selberg).

We are sending you a draft of the kind of statement which the nine have in mind to issue. The text is, of course, liable to minor changes; but must in the interest of expedience be considered substantially final.

Would you please let us know as soon as possible whether or not you wish your signature to be added?

Sincerely,
A. Pais.

P.S. Have your telegram. This text supersedes the previous one. If you agree, you need not cable again.

∽

---

[95]Courtesy of the Shelby White and Leon Levy Archives Center, Institute for Advanced Study, Princeton, NJ, USA.

Now that the official decision concerning the question of Dr. Oppenheimer's security clearance has been rendered, the undersigned permanent members and professors emeriti of the Institute for Advanced Study consider that in all propriety they may publicly express their feelings concerning Dr. Oppenheimer in the light of the charges brought against him.

We, who have known him as a colleague, as Director of our own institution, and as a neighbor in an intimate community, have complete confidence in his loyalty to the United States, his discretion in guarding its secrets, and his deep concern for its safety, strength, and welfare. We naturally therefore find satisfaction in the fact that by official decision his loyalty and patriotic devotion have been vindicated and his magnificent public service has been recognized.

Dr. Oppenheimer has performed for this country service of another kind, more indirect and less conspicuous but nevertheless, we believe, of great significance. For seven years now he has with inspired devotion directed the work of the Institute for Advanced Study, for which he has proved himself singularly well suited by the unique combination of his personality, his broad scientific interests, and his acute scholarship. We are proud to give public expression at this time of our loyal appreciation of the many benefits that we all derive from our association with him in this capacity.

THE INSTITUTE FOR ADVANCED STUDY
PRINCETON, NEW JERSEY

SCHOOL OF MATHEMATICS
June 18, 1954 [96]

Dr. George Placzek
Clarendon Physical Laboratory
Oxford, England

Dear George:

A few hours after the letter of June 17 was sent to you, von Neumann, Morse, Selberg, Montgomery and Goldstine agreed to sign the revised statement, a copy of which is herewith enclosed. All the others present will agree to this version for the sake of the unanimity that can thus be achieved. Please let us know as soon as possible whether or not you are willing to associate yourself with us in issuing this revised version in the event of an adverse decision by the Commission.

With best regards,

Harold F. Cherniss

P.S. If you agree, you need not cable again. -H.F.C.
P.P.S. Sorry that we have had so much trouble about this. -H.

☙

---

[96] Courtesy of the Shelby White and Leon Levy Archives Center, Institute for Advanced Study, Princeton, NJ, USA.

ATOMIC ENERGY OF CANADA LIMITED

CHALK RIVER PROJECT
CHALK RIVER, ONTARIO
CABLE ADDRESS "MOTA"

June 28, 1954[97]

Dr. G. Placzek
Institute for Advanced Study
Princeton, N.J., U.S.A.

Dear Dr. Placzek:

As you know, we have on file a large number of figures which were intended to form part of your report on the work of the Theoretical Physics Branch at Montreal. Dr. Weinberg asked me recently if this important material, which has long been eligible for declassification, could not be made available in some form. It is certainly a thing well worth doing but, unfortunately, we have no text of any kind to accompany the figures and only the barest hints on how to supply a text. Although, I am afraid, you have perhaps lost interest in this work, could you let us have some text, or materials from which to construct a text, to accompany the figures?

Henry H. Clayton
Branch Head,
Theoretical Physics

∞

---

[97]Courtesy of the Shelby White and Leon Levy Archives Center, Institute for Advanced Study, Princeton, NJ, USA.

Georg Placzek to Gordon Squires[98]

August 10, 1954

Dr. G. L. Squires
Atomic Energy Research Establishment
Harwell Nr. Didcot
Berkshire, England

Dear Dr. Squires:

May I trouble you about the following matter. During our first discussion at Oxford, I left with you a typewritten manuscript on *Neutron Scattering by Crystals*. I now find that this manuscript is needed for our work at Brookhaven. Accordingly, I should be most grateful if it were possible for you to send it to me at Brookhaven via air mail.

I am very much looking forward to your visit in Princeton and I hope you will let me know as soon as there are new developments concerning your fellowship or visa.

With best regards,

Yours sincerely,
G. Placzek

---

[98]Courtesy of the Shelby White and Leon Levy Archives Center, Institute for Advanced Study, Princeton, NJ, USA. Gordon Leslie Squires (1924-2010) was a Fellow of Trinity College, a constituent college of the University of Cambridge in England.

From G. Placzek to Robert Sachs[99]

Institute for Advanced Study
Princeton, NJ
March 2, 1955

Professor R.G. Sachs
Department of Physics
University of Wisconsin
Madison, Wisconsin

Dear Bob:

I finally managed to get hold of a reference to [Fritz] Houtermans' book:

> Beck and Godine, *Russian Purge*, Viking Press, New York.

Sincerely,

G. Placzek

Commentary: The exact title of this book is as follows:

> F. Beck and W. Godin, *Russian Purge and the Extraction of Confession*, (The Viking Press, New York, 1951). A remarkable story associated with this book is discussed in detail in *Physics in a Mad World*, Ed. M. Shifman (World Scientific, Singapore, 2015), pp. 267-288. F. Beck was Fritz Houtermans' pen name.

---

[99]Courtesy of the Shelby White and Leon Levy Archives Center, Institute for Advanced Study, Princeton, NJ, USA. Robert G. Sachs (1916-1999) was an American theoretical physicist, a founder and a director of the Argonne National Laboratory.

Léon Van Hove to Georg Placzek [100]

Prof. Dr. L. Van Hove
Instltuut voor Theoretlsche Natuurkude
Maliebaan 46. Utrecht
Telefoon 25 189 (K 3400)
March 3rd, 1955

Dr. G. Placzek
The Institute for Advanced Study
Princeton, N. J.

Dear George,

You will certainly have wondered why no news reached you in such a long time. It is partly due to the fact that we have been quite busy since New Year, partly to the fact that some difficulties occurred when I tried to confirm officially the date of my *oratie*. This has now been settled, but with a modification. It is namely on Monday June 6 that it will be held. I gathered from your letter of January 14 that this date would suit you as well as the previous one. Could you let me know whether this is so?

You asked me about my summer plans. As far as I can tell now, they will be very modest. I will be at Utrecht until the beginning of July, with perhaps one short interruption to attend the Pisa Conference on Elementary Particles from June 12 to June 17 (even this is not certain yet). In July and August I will probably be in Utrecht or Brussels or in some place near by. We are settled here since there is too short a time to travel great distances as soon as next summer. If you come to Heidelberg at the end of April, as you have written me, it would perhaps suit you to spend some time here in May. What do you think of this idea?

As I told you we have been fairly busy here in the last months. Van Kämpen has been appointed and is now settled in Utrecht. We have furthermore two American physicists, Foley [101] of Columbia University

---

[100] Courtesy of the Shelby White and Leon Levy Archives Center, Institute for Advanced Study, Princeton, NJ, USA.

[101] Henry Michael Foley (1917-1982) was an American experimental physicist, a leading researcher at Columbia University. In 1947, while a physics instructor at Columbia, he worked with Prof. Polykarp Kusch on measuring the magnetic moment of the electron, an achievement that in 1955 won the Nobel Prize in physics for Professor Kusch.

as a Fulbright professor and a younger man of the name of Brout,[102] also from Columbia. Nijboer is for the moment in Copenhagen. Now that the budget has been voted the last steps for his appointment as a professor can finally be taken (sickness of the Minister of Education had produced an additional delay). We have only one very serious difficulty remaining for the organization of the Institute here, namely that a plan to put a new building at our disposal failed in the very last stage. We are thus unexpectedly left with the fairly small space already available from the time of de Groot, although the appointments of new people are coming through at the expected rate. I guess it will still be some time before this unpleasant situation gets resolved.

As far as my own work is concerned I have made good headway during the winter. I have reconsidered the old problem of deriving quantum mechanical transport equations for systems where transport processes are produced by a small perturbation. From the practical cases of crystals one can isolate a special property of the perturbation which turns out to be entirely responsible for the irreversible effects. Using this property I gave a new derivation of the transport equation avoiding the main drawback of all conventional derivations, namely the necessity to assume random phases at all times. I have written up what I have done and I hope to send you a copy of the paper in a few weeks.

I also received your letter of February 16 concerning the misprints in our paper. The list you gave is impressive indeed and I am in a way happy not to have attempted to read the paper in print, since it would have only produced more worries about the number of mistakes. I do not think I will have the possibility to undertake a systematic hunt for further misprints in the immediate future, so that you might send the list as it stands to the editor. My private reprints arrived quite a while ago already.

With all my best greetings to everybody,

Yours very sincerely,
Lèon

---

[102]Robert Brout (1928-2011) was a Belgian theoretical physicist, a co-discoverer of the Higgs mechanism.

Georg Placzek to Victor Weisskopf[103]

March 3, 1955

Professor Victor Weisskopf
Department of Physics
MIT
Cambridge, Mass.

Dear Viki:

Enclosed herewith I am sending to you a copy of my letter to Mr. Moe. Many thanks for your help.

I am also returning to you with thanks the State Department booklet.

Sincerely,

G. Placzek

∽

Commentary: The nature of the booklet mentioned above becomes clear from the letter on page 219. It seems that Georg Placzek requested this pamphlet from Weisskopf because he was in the same situation as Weisskopf: namely, considering a job offer from Europe.

---

[103] Courtesy of the Shelby White and Leon Levy Archives Center, Institute for Advanced Study, Princeton, NJ, USA.

May 13, 1954

Prof. Victor M. Weisskopf
38 Arlington Street
Cambridge 36, Massachusetts

Dear Viki:

You have asked for an opinion as to the effect of the present immigration law upon your citizenship status for a possible acceptance of a permanent university appointment in England. You have indicated that your date of citizenship stems from May, 1943, that the proposed appointment commences in the autumn of 1955, and that it extends for an indeterminate period. In accordance with your wishes I have made a study of the present law, compared it with the old Nationality Act of 1940, and noted the relevant cases under each.

I call your attention to Chapter 3 of the present law, entitled "Loss of Nationality". In the interests of brevity I shall not enumerate the provisions of this chapter but refer you instead to pages 20-35 in the Department of State pamphlet dated December 24, 1952 and entitled "Information for Bearers of Passports".

Figure 3.4   The first page of a letter sent to Victor Weisskopf by his attorney, Gerlad A. Berlin, which was attached to Placzek's letter on page 218. This letter discusses the State Department pamphlet which had been sent to Weisskopf. From the first paragraph of Weisskopf's letter one can conclude that he was considering a job offer from England and possible legal consequences of its acceptance.

CONSEIL EUROPEEN POUR LA RECHERCHE NUCLEAIRE
EUROPEAN COUNCIL FOR NUCLEAR RESEARCH –
THEORETICAL STUDY GROUP

BLEGDAMSVEJ 15-17. COPENHAGEN – DENMARK
TELEGRAMS: PHYSICUM COPENHAGEN

TELEPHONE: TRIA 1616

March 30th, 1955 [104]

Professor George Placzek
The Institute for Advanced Study
Princeton, N. J.

Dear Placzek,

Rumors are spreading in Europe that you will be in Heidelberg at the end of April for the Kopfermann's birthday. Møller who – as you may know – is in charge of the Theoretical Study Division of CERN and who is at present at home with an attack of flu, has asked me to write you to enquire whether at this occasion, it would be possible for you to come to Copenhagen and give a lecture for the members of the Division. We should be very happy if you could pay us a visit, it might be before or after your visit to Heidelberg, as is convenient for you, and CERN would be glad to reimburse the expenses for your travel from Heidelberg to Copenhagen and back as well as for your stay here.

Looking forward to seeing you soon and with kind regards,

Yours sincerely,
Stefan Rozental

⌘

---

[104]Courtesy of the Shelby White and Leon Levy Archives Center, Institute for Advanced Study, Princeton, NJ, USA.

Bohr to Els Placzek [105]

December 31st, 1955

Dear Els,

Since the death of George I have wanted to write to you and say how much all his friends miss him and how deeply we sympathize with you in your loss. To me he was through many years not only a faithful friend, but whenever one met him his deep insight in scientific and human problems was a source of inspiration and encouragement. Ever since Aage and I in the wartime met you both in Montreal, I have understood how much you meant to George's life and Margrethe and I are grateful for the warm friendship yon have extended to both of us. We hope to meet you again before long and send you our warmest wishes.

Yours,

Niels Bohr

---

[105] Courtesy of the Shelby White and Leon Levy Archives Center, Institute for Advanced Study, Princeton, NJ, USA.

This letter was was written by L. Van Hove and addressed (tentatively) to the Institute of Advanced Study, carbon copy to Els Placzek.

Mrs. Els Placzek
Chalet Phoenix
Crans sur Sierre
Valais
Switzerland
Undated, after 1958 [106]

[...] As to the manuscript notes on the theory of Bose gases they concern a problem which Dr. Placzek wanted to study starting around 1954, and which would have occupied him during the academic year 1955-1956, which he planned to spend in Rome. The notes contain known equations concerning quantum statistics, and show that he had not really entered the new and original part of his research when illness and death prevented him from continuing. I enclose these notes for your historical archive, not because of the originality of their contents, but as an example of the methods of approach used by Dr. Placzek in his scientific investigations. In contrast to the neutrons notes, they are partly readable for outsiders.

I finally enclose a few papers relating to Dr. Placzek's intention to study the Vatican manuscript of the Geometry of Piero della Francesca. Dr. Placzek, among his many and rich intellectual interests, was a devoted student of the history of science and formed the plan to read in Rome Piero's works concerning geometry. I send you, for conservation in your archive, a letter of recommendation by Dr. E.A. Lowe, art historian at the Institute for Advanced Study, addressed to the conservator of the Vatican manuscripts, a description of the Piero manuscripts, and a list of references in Dr. Placzek's handwriting.

Yours sincerely, L. Van Hove

Director, Theory Division

∽

---

[106] Courtesy of the Shelby White and Leon Levy Archives Center, Institute for Advanced Study, Princeton, NJ, USA.

THE INSTITUTE FOR ADVANCED STUDY

PRINCETON, NEW JERSEY

Apr. 11/55

Dear Dom Albareda.

I take pleasure in presenting to you one of our most distinguished members, a professor of physics, whose reputation is international — Professor George Placezkitt is giving a course at the University of Rome, and at the same time he wishes to do

A letter of recommendation by Dr. E.A. Lowe, art historian at the Institute for Advanced Study, addressed to the curator of the Vatican Manuscripts Library. Courtesy of the Shelby White and Leon Levy Archives Center, Institute for Advanced Study, Princeton, NJ, USA.

some research. He is parti-
cularly eager to examine the
Vatican manuscript on
Geometry by Piero della Francesca.

Will you kindly make that
volume available to him and
give him all the facilities you
are well known to give to all
serious investigators.

I wish I were accompany-
ing Prof Placzek, for I should
then have the chance of seeing you again & recall old
times. My respectful & affection-
ate greetings to Cardinal Mercati.

Sincerely yours, E. A. Lowe

A letter of recommendation by Dr. E.A. Lowe, art historian at the Institute for Advanced Study, addressed to the curator of the Vatican Manuscripts Library, page 2.

George Placzek to Albert Andriesse

Zurich
September 25, 1955

Albert Andriesse,
Scarsdale, NY

Dear Albert,

Again my thoughts are very much with you. I know how little con-solation words can convey, but perhaps you can find it in the certitude that you have made things as bearable for poor Ella as possible and surrounded her with love in every way.

It is with a heavy heart that I see Els leave and I hope she will be able to bring you some solace. I feel rather ashamed to be unable to come with her which would have been the best solution in every respect, but it really does not work, not because I do not want to, but for reasons which I hope you will understand; this is certainly not the time to entertain you of my troubles.

Yours as ever,

George Placzek

⌘

Commentary: This is the last letter of George Placzek of which we are aware. It is written to his father-in-law, Albert Andriesse, in connection with the death of Placzek's mother-in-law, Ella Andriesse-van den Bergh, that same day. Credit: Andriesse Family Archive, courtesy of Ella (Elisabeth) Andriesse, New York.

See also page 54.

László Tisza to Aleš Gottvald

July 23, 2003

I have answers to a few of your questions:

1) I did meet G. Placzek in Kharkov, probably in 1936. This, I believe was our first meeting in person, but we had significant indirect contacts before. Placzek developed a beautiful theory of the Raman effect. He reviewed the theory in a volume of the *Leipziger Vorträge*, where he discussed the selection rules in the Raman spectra of symmetric polyatomic molecules. Like so many physicists at the time he didn't like group theory and his intuitive method was not quite reliable. Edward Teller was in touch with both of us and suggested that I develop the group theory of the selection rules. This became my PhD thesis (*Z. f. Physik*, 82, 48 (1933)). Placzek was not only an excellent physicist, he had also a sharp satirical tongue. Reminds me of Guido Beck of similar background.

2) I met Placzek again in September 1937 in Paris at the "Congrès du Palais de la Découverte." There were many social events and one of the stars was the wife of Hans von Halban, Jr. who worked with Fred Joliot. Placzek did not conceal his admiration for her. Shortly after, the Halbans got a divorce and Placzek married her. I know nothing more about him.

Yours,

Tisza

∽

Michael Vinegrad to M. Shifman

Central Archive
of Hebrew University of Jerusalem
Michael Vinegrad
michaelv@savion.huji.ac.il

Mikhail Shifman
University of Minnesota
July 24/2016.

Dear Professor Shifman,

The Central Archives of the Hebrew University has Placzek's personnel file. The documents in the file relate to the conditions of his employment.

From the December 31st letter to the University's Chancellor Dr. J.L. Magnes (enclosed) it seems he'd arrived in Jerusalem by the end of December 1934 but no document mentions his exact arrival date.

The Assistant Registrar M. Ben-David certified on January 9th, 1935 that "Professor Dr. Georg Placzek is Guest-Professor in Theoretical Physics at this University" and on January 28th, 1935 that "Professor G. PLACZEK is Guest-Professor at the Hebrew University." On August 13, 1935, Professor Placzek wrote to Dr. Magnes (letter enclosed) and on August 15th M. Ben-David wrote to Professor Placzek that his [last] salary would be paid on September 30, 1935.

On 26.XII.38 Secretary M. Ben-David certified that "Professor G. Placzek was guest Professor of the Hebrew University from December 1934 to September 1935."

Your sincerely,

Michael Vinegrad

∽

Figure 3.5   The Hebrew University of Jerusalem in 1932 (Courtesy of Michael Vinegrad).

PLACZEK'S CURRICULUM VITAE (VARIOUS YEARS)

Chapter 4

Placzek's Curriculum Vitae (Various Years)

"Above the fireplace was a large mirror on which Placzek is reported to have customary written mathematical equations with lipstick or soap"

From the FBI Klaus Fuchs file, 1950

Georg Placzek to Chancellor Dr. J.L. Magnes [1]

THE HEBREW UNIVERSITY

Stamp: January 13, 1935

Jerusalem

December 31, 1934

Dear Dr. Magnes:

I will take the liberty to give you a short overview of my scientific positions with reference to our conversation of yesterday.

From January to May 1931 I worked at the Theoretical Physics Institute of the Technical College of Stuttgart under the leadership of Professor P. P. Ewald. Then my research topic was, among others, the crystal structure theory, my main scientific contacts were Mr. P. P. Ewald,[2] Carl Hermann and Martin Ruhemann.

From May to August 1931 I was employed at the Theoretical Physics Institute of Leipzig University where I could work with the heads of the Institute, Professor Werner Heisenberg and Professor Friedrich Hund, as well as with a number of younger physicists. The problems we worked

---

[1]Courtesy of the Central Archive of The Hebrew University of Jerusalem. See also page 233.

[2]Professor Paul Peter Ewald (1888-1985) was a German crystallographer and physicist, a pioneer of X-ray diffraction methods.

231

on concerned the newer quantum theory, especially the theory of molecular structure.

In July 1931 I was a lecturer in the "Leipziger Woche" (Leipzig Week), an annual discussion series about special topics of physics, upon the invitation of Professor Peter Debye.

From October 1931 to June 1932 I was employed at the Theoretical Physics Institute of Rome University where I cooperated in experimentation with Eduardo Amaldi and in theoretical research with Enrico Fermi, Hans Bethe and Edward Teller researching in the above-mentioned domains.

In July and August 1932 I worked with Edward Teller in Göttingen.

In September 1932 I participated in the German Physicists' Society conference in Bad Ems[3] and delivered a lecture there.

From September 1932 to April 1933 I worked at the Theoretical Physics Institute of Copenhagen University under the leadership of Prof. Niels Bohr. In February 1933 I visited the Theoretical Physics Institute at Leipzig University (see above) for a short time.

From May to November 1933 I was a guest at the Ukrainian Institute for Physics and Technology in Kharkov,[4] which belongs to the People's Commissariat of Heavy Industry. Then I came back to Copenhagen where I have been working until today.

Should you wish to learn other details, I will of course always be at your disposal.

With the highest esteem and regard

Yours faithfully,
G. Placzek

∽

---

[3] Bad Ems is a town in Rheinland Pfalz, West Germany.

[4] The second of Placzek's visits to Kharkov took place from early December 1936 till mid-January 1937. It lasted for about six weeks.

Sehr geehrter Herr Dr.Magnes,

    Im Anschluss an unser gestriges Gespräch
erlaube ich mir,im Folgenden eine kurze Uebersicht über meine
wissenschaftliche Tätigkeit ~~zusgeben~~ in den letzten Jahren zu geben.
    Im Jahre <u>1931</u> war ich am Institut für theoretische
Physik der Technischen Hochschule Stuttgart (Leitung Prof.P.P.
Ewald) von Januar bis Mai tätig.~~und~~ Ich beschäftigte mich damals
u.a. mit Problemen der Theorie des Kristallbaus,wobei ich haupt-
sächlich mit den Herren P.P.Ewald, C.Herrmann und M.Ruhemann wis-
senschaftlichen Kontakt hatte.

    Vom Mai bis August 1931 arbeitete ich am Institut für theo-
retische Physik der Universität Leipzig, wo ich mit den Leitern des
Instituts,Prof.W.Heisenberg und Prof.F.Hund, sowie mit einer Reihe
jüngerer Physiker zusammenarbeiten konnte. Die bearbeiteten Fragen
betrafen die neuere Quantentheorie,insbesondere die Theorie des
Molekülbaus.

    Im Juli 1931 nahm ich auf Einladung von Prof.P.Debye als Vor-
tragender an der"Leipziger Woche" -eine Diskussionsreihe über spezi-
elle physikalische Themen, die alljährlich stattzufinden pflegt-teil.

    Von Oktober 1931 bis Juni <u>1932</u> arbeitete ich am physikali-
schen Institut der Universität Rom ,wobei ich experimentell mit
E.Amaldi und theoretisch mit E.Fermi,H.Bethe und E.Teller über
Probleme dergenannten Forschungsgebiete zusammenarbeitete.

    Im Juli und August 1932 war ich in Göttingen und arbeitete
dort zusammen mit E.Teller.

    Im September 1932 nahm ich an der Tagung der Deutschen Physi-

The first page of the original of the letter on page 231.

[George Placzek's] CV

Undated, presumably written in Montreal around 1942-43. [5]

Born 1905

Dr. Phil., Vienna 1928;

Thesis: *On determination of densities and shape of submicroscopic test bodies (experiments in inhomogeneous electric field).*[6]

1928-1932: Research work on experimental spectra and atomic physics at Physics [Institute] 1 of the University of Utrecht (L.S. Ornstein, H.A. Kramers);

Stutgart (T. H. P.)

Leipzig (W. Heisenberg)

Rome (E. Fermi)

1932-1933, 1934 (June-September), 1935-1938:
Scientific Collaborator at the Institute for Theoretical Physics in Copenhagen (N. Bohr). Part of that time as Oersted Fellow.

1933 (May-December): Lecturer and Research Consultant with Ukrainian Physical-Technical Institute, Kharkov, USSR.

1934-1935: Professor of Theoretical Physics at the University of Jerusalem.

After 1939: Research Assistant at Cornell

Teaching

Duties at Institute for Theoretical Physics, Copenhagen:
Directing research work of scientific workers and graduate students;

Ukrainian Physical-Technical Institute:
Courses for graduate students and research workers on molecular theory, quantum mechanics and nuclear theory;

University of Jerusalem:
Lectures on Theoretical Physics and direction of seminars mostly for graduate students;

Institut de Chimie Nucleaire, Collège de France, Paris:
Lectures on Nuclear Physics for graduate students and research workers;

---

[5]A handwritten draft on unnumbered pages, some of them in French. See page 67 for the first page of the original. George Placzek was among the first arrivals at the Montreal Laboratory as the head of their theoretical division. He and Hans von Halban who headed Montreal Laboratory arrived at the end of 1942. Courtesy of AIP.

[6]See page 253.

Cornell University:
Courses for graduate students (theory of solids, atomic and molecular theory, quantum mechanics). Direction of seminars. Supervision of the work of summer lecturers and of graduate students.

Work

Work for a number of years with Prof. Niels Bohr on general quantum mechanics, the theory of radiation and the theory of the mechanisms of nuclear reactions. My own work till 1936 was mostly in the fields mentioned above and after 1936 mostly in nuclear theory. For the last four years I have been particularly engaged in work on the theory of neutron diffusion.

Research in experimental physics

Electricity and colloid physics (PhD thesis, Vienna 1928);
Spectroscopy, measurements of intensity and polarization of spectral lines (Utrecht 1928-1930; Rome 1931-1932);
Detection and properties of slow neutrons (Copenhagen 1936).

MANHATTAN DISTRICT

SCIENTIFIC RESEARCH & DEVELOPMENT PERSONNEL

(Please PRINT or TYPE. Answer all questions fully. If "None", so indicate. If space is insufficient attach additional sheets as necessary.)

1. NAME: PLACZEK   George
   Last   First   Middle Initial

2. PRESENT ADDRESS: Box 1663   Santa Fe   New Mexico
     No. and Street   City   State

3. PLACE WHICH YOU REGARD AS YOUR "HOME TOWN": New York, New York
             City   State

4. PLACE OF BIRTH: (Bruenn, Austria, since 1918 Czechoslovakia) DATE OF BIRTH: Sept. 26, 1905
   City   State or Country   Month Day Year

6. FAMILY: List the following members of your family, even though deceased:

	NAME	HOME ADDRESS (City & State)	OCCUPATION
FATHER	Alfred	deceased	
MOTHER	Marianne	unknown	
SPOUSE	Fanny Ella	Box 1663, Santa Fe, N.M.	
BROTHERS (B)	(B) Frederic	deceased	
SISTERS (S)	(S) Edith C.	unknown	
CHILDREN (C)			
(Designate after each name B, S, or C)			

FINAL DETERMINATION
UNCLASSIFIED
L. M. Redman
FEB 11 1981

7. MILITARY SERVICE: List your military service and that of members of your family listed above:

NAME	BRANCH OF SERVICE	RANK	YEARS FROM	TO
None				

MD-H-1

The first page of George Placzek's personnel questionnaire form from Los Alamos, June 22, 1945 (Courtesy of Alan Carr, Bradbury Science Museum, LANL). From other pages we learn, in particular, that Placzek was a member of École Libres des Hautes Études in New York since 1942. The latter was a sort of university-in-exile for French academics in New York during the Second World War.

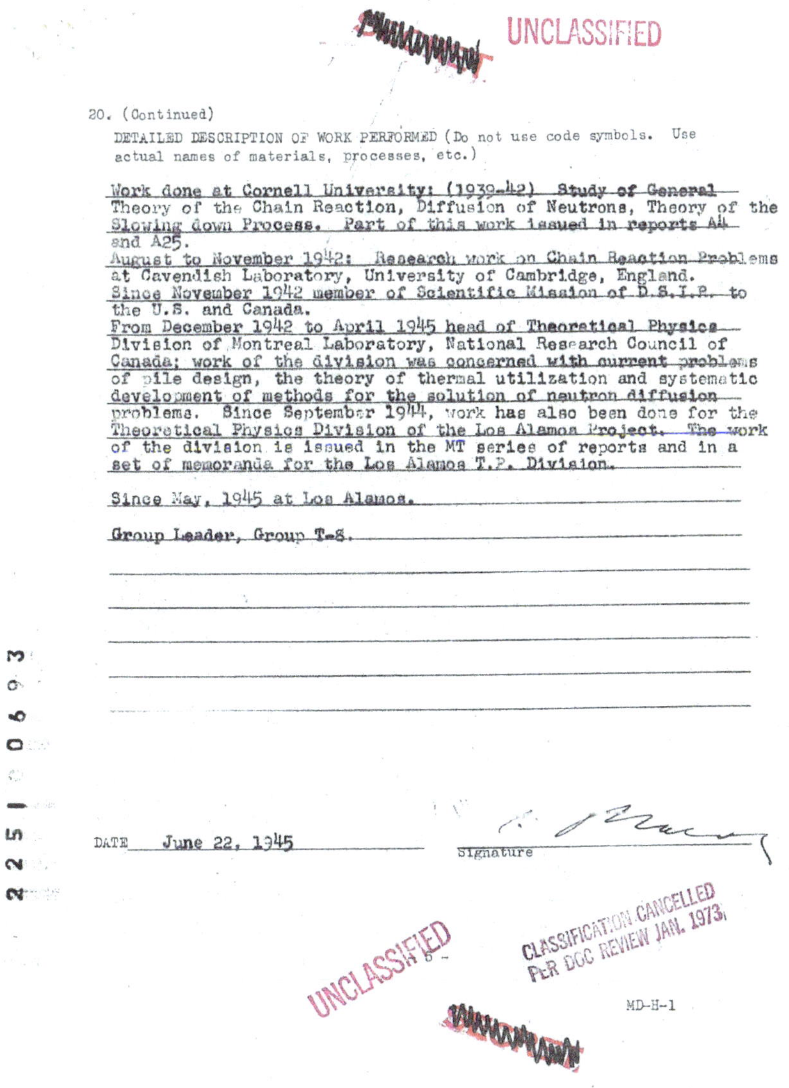

George Placzek's personnel questionnaire form (the last page) from Los Alamos, June 22, 1945 (Courtesy of Alan Carr, Bradbury Science Museum, LANL).

Publications by G. Placzek[7]
(undated, presumably mid-1948)

1) "Ueber ponderomotoriwohe Wirkungen des Lichts auf ungeladene submikroskopische Körper im elektrischen Felde," *Zs. f. Phys.* 49, 601, 1928.

2) "Zur Dichten und Gestaltsbestimmung submikroskopischer Probekorper," *Zs. f. Phys.* 55, 81, 1929.

3) "Zur Theorie des Ramaneffektes," *Zs. f. Phys.* 58, 585, 1929.

4) "Ueber den Ramaneffekt beim kritischen Punkt," Proc. Amsterdam, 33, 832, 1930.

5) "Ueber die Lichtzerstreuung beim kritischen Punkt.," *Phys. Zs.* 31, 1052, 1930.

6) "Polarisationsmessungen am Ramaneffekt von Flüssigkeitem," (With W. R. van Wijk), *Zs. f. Phys.* 67, 582, 1930.

7) "Intensitaet und Polarisation der Ramanschen Streustrahlung Mehratomiger Molekuele," *Zs. f. Phys.* 70, 287, 1931.

8) "Ueber das Kontinuierliche Ramanspektrum und sein Verhalten beim Kritischen Punkt.„ (With W. R. van Wijk), *Zs. f. Phys.* 70, 84, 1931.

9) *Ramaneffekt und Molekuelsymmetrie*, (Leipziger Vortraege, S. Hirzel, 1931), page 71.

9a) *The structure of molecules*, Ed. by P. Debye. London, Blackie and Sons, 1932 (English edition of (9)).

10) "Evidence for the spin of the photon from light scattering," *Nature*, 128, 410, 1931.

11) "Ramaneffekt des gasfoermigen Ammoniaks," (With E. Amaldi), *Neturwiss.* 20, 521, 1932.

12) "Die Rotationsstruktur der Ramanbanden mehratomiger tlolokuele," (With E. Teller), *Zs. f. Phys.* 81, 259, 1933.

13) "Ueber das Ramanspektrum des gasfrmigen Ammonisks," (With E. Amaldi), *Zs. f. Phys.* 81, 254, 1933.

14) "Rayleighstreuung und Raman effekt," *Handbuch der Radiologie*, Ed. E. Man, vol. VI/2, Leipzig, 1934, page 209.

15) "Struktur der unverschobenen Streulinie," (With L. Landau), *Physikalische Zeitschrift der Sowjetunion*, 5, 172, 1934.

16) *Quantum theory of Light Scattering*, (Monograph in Russian), Kharkov, Editions ONTI, 1935.[8]

---

[7]Courtesy of the NRC Archives, Canada.
[8]More exactly, Релеевское Рассеяние и Раман-Еффект [Releyevskoe Rasseyaniye i Raman

17) "Capture of slow neutrons," (With O. R. Frisch), *Nature*, 137, 357, 1936.

18) Resonance Effects in Nuclear Processes. (With H. A. Bethe). *Phys. Rev.* 51, 450, 1937.

19) "La Spectroscopie der Neutrons lent," Lectures held at the Institut Henri Poincare in 1938. *Annales de l'Institut Henri Poincare*, in course of publication.

20) "Slowing down of neutrons by heavy nuclei," *Phys. Rev.* 55. 1130, 1939.

21) "Nuclear Reactions in the Continuous Energy Region," (With N. Bohr and R. Peierls), *Nature*, 144, 200, 1939.

22) "On the Interpretation of Neutron Measurements in Cosmic Radiation," (With H. A. Bethe and S. A. Korff), *Phys. Rev.* 57, 573, 1940.

23) [left blank]

24) "The Spatial Distribution of Neutrons Slowed Down by Elastic Collisions," Declassified Manhattan District Report A-25, 1941.

25) "The Functions $E_n = \int_1^\infty \exp(-xu)u^{-n}du$," Document MT-1, National Research Council of Canada, Division of Atomic Energy, 1943.[9]

26) "Diffusion of Neutrons without change in Energy," (With G. M. Volkoff), Document MT-4, National Research Council of Canada, Division of Atomic Energy, 1943. [See footnote 6 on page 238.]

27) "The Neutron Density near a Plane Surface," Document MT-16, NRC, DAC, 1943. [See footnote 6 on page 238.]

28) "On the Theory of the Slowing Down of Neutrons in Heavy Substances," *Phys. Rev.* 69, 473, 1946.

29) "The Angular Distribution of Neutrons Emerging from a Plane Surface," *Phys. Rev.* 72, 556, 1947.

30) "Milne's Problem in Transport Theory," (with W. Seidel), *Phys. Rev.* 72, 550, 1947.

31) "A Theorem on Neutron Multiplication," (with G. M. Volkoff), *Can. Journ. of Research*, A25, 276, 1947.

32) "The Concept of Albedo in Elementary Diffusion Theory," in: *The Science and Engineering of Nuclear Power*, Vol. 2, Cambridge, 1948.

33) "The Mechanism of Nuclear Reactions," (with N. Bohr and

Effect, Nauchno-Tekhinicheskoe Izdatelstvo, Kharkov, 1935]. See entry (14) above.

[9]Obtainable from Plans and Publications Division, Ottawa, Canada. -G. Placzek

R. Peierls), Proceedings of Copenhagen Academy, in press.[10]

Commentary 1:   Placzek did not include in the above list two Montreal Laboratory reports, presumably because they were not declassified by the time of compilation of this document.

MT-51: *Energy conversion factors and nuclear constants,* (completed February 2nd, 1943, issued April 24th, 1944, 2 pp.)

MT-118: *Milne's problem with capture and production,* (with B. Davison, issued March 5th, 1945, 16 pp.)

Commentary 2:   We believe it would be convenient to supplement the above list with Placzek's works published after 1948.

[a] "Theory of Slow Neutron Scattering," *Phys. Rev.*, 75, 1295, 1949.

[b] "Effect of Short Wavelength Interference on Neutron Scattering by Dense Systems of Heavy Nuclei," (with B. Nijboer and L. Van Hove), *Phys. Rev.*, 82, 392, 1951.

[c] "Correlation of Position for the Ideal Quantum Gas," Proceedings of the Second Berkeley Symposium on Mathematical Statistics and Probability, 1950: 581-588 (1951).

[d] "The scattering of neutrons by Systems of Heavy Nuclei," *Phys. Rev.*, 86, 377-388 (1952).

[e] "Scattering of X-Rays by Atoms," *Phys. Rev.*, (2) 86, 588, 1952.

[f] *Introduction to the Theory of Neutron Diffusion*, vol. I, (with K. M. Case and F. de Hoffmann), Los Alamos, Los Alamos Scientific Laboratory, 1953, 174 pp.

[g] "Crystal Dynamics and Inelastic Scattering of Neutron," (with L. Van Hove), *Phys. Rev.*, (2), 93: 1207-1214, 1954.

[h] "Incoherent Neutron Scattering by Polycrystals," *Phys. Rev.*, (2) 93, 895-896, 1954.

[i] "Interference Effects in the Total Neutron Scattering Cross-Section of Crystals," (with L. Van Hove), *Il Nuovo Cimento*, (1), 233-256, 1955.

[j] *The Rayleigh and Raman Scattering*, translated by Ann Werbin from entry (14) above, (Berkeley, CA. 1959).

---

[10]This paper never appeared, for details see page 31.

PART 3

PLACZEK IN THE LITERATURE, MEDIA, AND RECOLLECTIONS. OBITUARIES

Placzek in the Literature, Media, and Recollections. Obituaries

Excerpts from
The Genesis of Physics at the Hebrew University of Jerusalem
by Issachar Unna
Phys. Perspect., **2**, 336 (2000)

The George Placzek Affair: The Theoretician who Demanded a Laboratory

By 1934 a suitable appointment to the chair of theoretical physics still had not been made, even though a number of first-class candidates had become available. Ornstein, in a letter of March 1934 to Chancellor Magnes, proposed three names: George Placzek ("who is at present in Copenhagen"), Walter Heitler ("at the moment in Bristol"), and Lothar Nordheim ("who is said to come to Utrecht"). Placzek, heading the list, was only 28 years old, but already known as an incisive, brilliant physicist of broad education and a fascinating personality. As it turned out, he actually stayed at the Hebrew University for six months during the academic year 1934-35. The process of his appointment and the way it was dealt with during his stay in the country – until he "slammed the door" on the university – were characteristic of the difficulties and complications that accompanied this early stage in the development of the physics department.

George Placzek (1905-1955) was the first-born of a wealthy Jewish family in Brünn, Moravia (today, Brno, Czech Republic). His talents in science and languages already stood out during his years in secondary school, when he became well known for his broad and deep erudition. He pursued his higher education at the Universities of Prague and Vienna, where he received his doctorate in 1928. His ever-increasing circle of friends and colleagues in Europe found him to be a charismatic, witty man with a deep knowledge in physics. From 1928 to 1931 he worked in Utrecht with H. A. Kramers and Ornstein; from 1931 to 1932 he was with Enrico Fermi in Rome, where he also met Hans Bethe, Edward Teller, and Edoardo Amaldi; and from 1932 to 1934 he was with Niels Bohr, Otto Robert Frisch and others in Copenhagen. In 1934 he also worked for six months with Lev Landau in Kharkov.

When Ornstein proposed Placzek for a post at the Hebrew University in 1934, he was already well known owing to fundamental articles that he had written on the theory of Raman scattering. In 1933 he published jointly with Teller an article on "The rotational structure of Raman bands in multi-atomic molecules," in which they used the methods of group theory and methods that later were recognized as part of Wigner's and Racah's methods in spectroscopy. Another article, published in collaboration with Amaldi, was an experimental work confirming the above theoretical calculations. In 1932, after the discovery of the neutron, Placzek immediately turned his attention to nuclear physics and within a few years was recognized as an authority on the scattering of neutrons and their absorption in matter. He was one of the first to recognize the importance of and contribute to the new field of nuclear physics.

In June 1934, Sambursky asked Magnes to speak to Placzek, who was in Berlin, and to convince him to come to the Hebrew University for more than half a year. Ornstein, too, wrote to Placzek in August telling him that the language of instruction in Jerusalem is Hebrew, but that one is permitted to teach in another language for one year. The Board of Governors of the University met later that month in Zurich with the subject of physics on its agenda. The Board adopted the following two resolutions: (1) to appoint Sambursky as lecturer in experimental physics; to confirm Alexander as permanent assistant; and to appoint Wolfson as permanent assistant at the physics institute; and (2) to found, at the physics institute, a chair for theoretical physics and to empower the executive committee to appoint the professor to this chair, subject to the endorsement of the university council. Four candidates for this chair were proposed, to be approached in the following order: Placzek, Heitler, London, Nordheim.

The discussions at the university about Placzek's nomination were difficult and agitated, and continued until November 1934. A special committee was appointed whose members were A. H. Fraenkel, M.

Fekete, A. Fodor, S. Klein and S. Sambursky. The committee expressed concern that the award of a professorship to someone so young might cause discontent among the rest of the faculty of the university, and it therefore proposed to accord him the rank of lecturer and to elevate him to professor only after he had become proficient in Hebrew. Ornstein, who was charged to negotiate with Placzek, reacted with chagrin to this proposal: he did not agree with the committee and argued that its considerations were miserly; that its members did not understand the difficulties in finding a first-rate theoretical physicist ("that is how they lost Felix Bloch"); and Placzek was an eminent one. Ornstein was prepared to approach Placzek only if the committee agreed to accept him at the rank of professor, and for one full year. Meanwhile, a financial problem arose: the special fund to be used for the appointment was designated for refugee scientists from Germany, and Placzek was not a refugee from Germany! Salman Schocken, Chairman of the Executive Committee, blamed Ornstein for overlooking the aim of appointing a refugee from Germany, but Schoken also understood from a conversation with Weizmann that Weizmann "would see any interference with Placzek's invitation as a serious blow to the University." In May 1935, a meeting took place in Chancellor Magnes's office, and in the presence of S. Ginzberg, the Academic Secretary, Magnes discussed with Placzek his terms for accepting a permanent appointment in Jerusalem. Placzek's terms included personal items, but most of his conditions related to the development of the physics department. He maintained that Ornstein had promised him a travel budget to Europe (200 P).[1] Magnes explained that "we do not have such an arrangement, and there is no hint as to it in Ornstein's letter." Placzek said that Ornstein had promised him a high-level assistant, a position he would fill with Edward Teller. Magnes did not object, probably because he understood from Placzek that this in any case was not feasible in the near future. The discussion then expanded to the place of theoretical physics in general in the Hebrew University, and Placzek promised to prepare a memorandum on this question. After the meeting, Magnes reported: "I was disappointed to hear that Placzek thinks an investment of 20,000 P is needed, and a yearly expenditure of 3,500 P, but we were told that a theoretical physicist needed only a blackboard and chalk or paper and pencil."

---

[1] The Palestine pound was the currency of the British Mandate of Palestine from 1927 to May 14, 1948. It was introduced as an equivalent to the British pound. 1P in 1934 is roughly equivalent to $100 in 2017.

Placzek insisted in his meeting with Magnes on the foundation of a department of experimental nuclear physics – this only three years after the discovery of the neutron. Magnes commented that he intended to assist only the laboratories for optics and physical chemistry. Placzek answered that others might be content with that, but he did not expect from Ornstein that these laboratories actually contained such poor equipment. "Here, the equipment is inferior to that of a secondary school in Germany. This is not criticism. It is only a description of the situation." To improve this situation, because Placzek continued to maintain his great interest in moving to Palestine, he proposed to turn to Arthur H. Compton in Chicago, James Franck at The Johns Hopkins University in Baltimore, and to Einstein for help in raising funds, for otherwise it will be impossible to act.

Placzek was willing to postpone a discussion of his academic rank, but at the same time he advanced definite administrative demands: a separate secretariat for physics and physical chemistry, and separate and less cumbersome and complicated management ("desks which had been ordered months ago arrived only now"). Magnes commented that even a separate secretariat would not expedite such administrative matters. Then the question of an assistant was raised again (Teller?), and here Magnes noted: "this will require so many committees..." At this point M. Schloessinger (the deputy chancellor) entered the room, and asked Placzek if he had another offer of an appointment somewhere else, to which Placzek answered: "at this moment this is beside the point because I would like to stay here."

A week later S. Ginzberg, the Academic Secretary, sent a letter to Magnes in which he discussed Placzek's demands. Concerning a separate secretariat for physics and physical chemistry, Ginzberg voiced a negative opinion. Placzek's demands regarding a physics laboratory also were greatly exaggerated; he should not be allowed to recruit funds in America. Placzek's demands for his salary and traveling expenses were totally out of bounds; in Ginzberg's words, "he himself said that maybe somebody else would be content with less." Ginzberg concluded that Placzek and the university were incompatible, and "the sooner we decide [this] the better." They would have to find someone else, and Ginzberg then raised the possible candidacy of Felix Bloch: "I heard that Dr. Bloch could be taken into account. From what one hears he is a better physicist, and a better Jew."

Schloessinger, too, sent a note to Magnes saying: "His [Placzek's] demands are so exaggerated as to make it impossible for me to take the whole matter seriously. Also, I do not recommend writing about it to America, as this would reflect unfavorably on the acumen of our judgment."

In June 1935, Placzek submitted two documents to Magnes specifying his demands. The first, on June 5, dealt with theoretical physics, and here he again advanced the candidacy of Edward Teller as assistant, although by then it was clear that Teller could not come to Jerusalem for at least a year or two, since he had just received an appointment at George Washington University. The second document, on June 15, was a detailed proposal for the foundation of a nuclear physics department, with a budget of 27,000 P.

Meanwhile, however, even before Placzek submitted his second document to Magnes, Magnes himself had sent a letter on June 10 to Schloessinger and Schocken, in which he stated unequivocally: we cannot agree to Placzek's demands, and we therefore must send a telegram to Bloch in California opening negotiations with him. Almost simultaneously, Schocken sent a letter to Magnes, in which he did not foreclose the possibility of accommodating Placzek's demands, mainly because "he is ready to start teaching in Hebrew next year." Soon thereafter, taking Placzek's documents and everything else into account, Magnes sent Placzek an official letter, dated July 4, 1935, in which he offered Placzek an appointment to a permanent position at the university under the following conditions:

> Position of lecturer with rights of professor at a salary of 420 P per annum, "the highest salary at the university."
>
> The sum of 88 P, once, for travel.
>
> Assistance – one orderly assistant.
>
> No additional secretarial assistance will be provided.
>
> No possibility of nuclear physics (experimental) research will be provided.
>
> No promise of additional rooms.

One week later, on July 11, another meeting took place between Placzek and Magnes. Magnes wrote (in Hebrew!) a memorandum of that meeting, reporting:

> Was with me for an hour. Announced that he could not accept the conditions of appointment offered him in the letter of July 4, 1935. He wanted to participate in building up, and we offer him static conditions. There were

too many misunderstandings from the outset, but he does
not want to blame Ornstein for this. We stand no chance
in nuclear physics and therefore he is right to leave, although he could work 5 months (July-December) each
year abroad. Magnes, however, thought that these were
not "static" conditions but first steps toward the creation
of theoretical physics at the university. "After all, he is
still young... [What] bothers him is the tendency [Einstel-
lung]. [He thought] conditions would be more amenable.
He is sorry, because the country made a deep impression
on him. Was far from Judaism but the country began to
fascinate him [Ihn zu faszinieren].

I repeated my offer if he could come to terms with the fact that there
is no chance for nuclear physics. Answered it was impossible. Would
help as much as he could. Spoke about Bloch. To his mind – Bloch
will not accept our conditions. He mentioned other names: Wigner in
Princeton, Peierls from Munich (now in Manchester), Breslau – older,
Fröhlich – a young man who was in Russia.[2] Placzek himself plans
to go to England, and afterwards to Copenhagen and possibly also
to America. Does not believe it would be difficult to get a suitable
appointment.

A couple of days later, Placzek sent an official letter to Magnes,
giving his reasons, as above, for rejecting the offer. His departure was
accompanied by another discord, when it turned out that he did not
receive the entire salary that the university had promised him in a
telegram to Ornstein in November 1934: He received only 300 P instead of 400 P. Someone in the academic secretariat claimed that the
promised amount was for the entire academic year whereas Placzek
actually stayed in the country only six months.

Placzek returned to Copenhagen after sending a telegram to Otto
Robert Frisch: "THROUGH WITH JEWS FOR EVER." In Copenhagen, Placzek made significant contributions to nuclear physics, becoming an authority on the scattering and absorption of neutrons, and
nuclear fission. He used Bohr's Nobel Prize medal, for example, to
measure the absorption cross section for slow neutrons in gold. He
stimulated Frisch to carry out the first experiments that confirmed

---

[2]Herbert Fröhlich (1905-1991) lost his position of *Privatdozent* at the University of Freiburg in
1933. That same year Fröhlich went to the Soviet Union at the invitation of Yakov Frenkel, to work
at the Ioffe Physical-Technical Institute in Leningrad. His Soviet visa was abruptly terminated in
1935, and he had to flee to England immediately.

nuclear fission. He left Copenhagen for Paris in 1938 and in 1939 was appointed as a research associate at Cornell University. In 1942 he became head of the department of theoretical physics at the Chalk River nuclear research center in Canada. At the end of World War II, he moved to Los Alamos for a year, then became a research physicist in the General Electric Company before being appointed member of the Institute for Advanced Study in Princeton in 1948. He died in October 1955 of a heart attack, at the age of fifty.

George Placzek – an Unsung Hero of Physics[3]

Jan Fischer

On 21-24 September, in cooperation with several other scientific institutions, the Masaryk University in Brno is organizing a memorial symposium in honor of the Czech physicist George Placzek, who was born on 26 September 1905 and died on 9 October 1955, soon after his 50th birthday. Placzek was an outstanding scientist who made substantial contributions to the fields of molecular physics, scattering of light from liquids and gases, the theory of the atomic nucleus and the interaction of neutrons with condensed matter. He belongs among the most important physicists of the 20th century, setting an example not only through his discoveries, but also by the stimulating style of his scientific work.

George Placzek was born in Brno, Moravia, in what is now the Czech Republic but which in 1905 was part of Austro-Hungary. The oldest son of Alfred and Marianne Placzek, he spent his childhood in Brno and in Alexovice, where the family owned a textile factory, Skene & Co. He had a brother, Friedrich, one year younger and a sister, Edith, 12 years younger. The family was well integrated into the mixed Czech-German language environment around Brno. George studied in the Deutsches Staatsgymnasium in Brno between 1918 and 1924 and then went to the University of Vienna, with three semesters away in Charles University, Prague. He graduated in 1928, having defended his doctoral thesis, which dealt with the determination of density and shape of submicroscopic test bodies, with distinction.

Travels in Europe

In 1928, Placzek set off on several years of travel through the main scientific centers of Europe. This was usual for post-docs at the time, although Placzek later markedly avoided countries in which Adolf Hitler was increasingly encroaching upon civil liberties and human rights. He spent three years with Hendrik Kramers in Utrecht, followed in 1930-31 by a short time with Peter Debye and Werner Heisenberg in Leipzig. He then joined a group of young physicists led by Enrico Fermi in Rome, where Edoardo Amaldi became his closest colleague. Then, in 1932, Placzek joined Niels Bohr in Copenhagen where he remained until

---

1938, with periods of research fellowships or visiting professorships at the universities of Kharkov, Jerusalem, Paris and elsewhere.

Placzek's first scientific interest was in the scattering of light from molecules and the Raman effect. With Lev Landau, he investigated the fine structure of a monochromatic wave in liquids and gases, and together they derived the Landau-Placzek formula for the ratio of intensities of Brillouin and Rayleigh scattering of light. Then in the early 1930s, the scattering of slow neutrons in matter became topical and

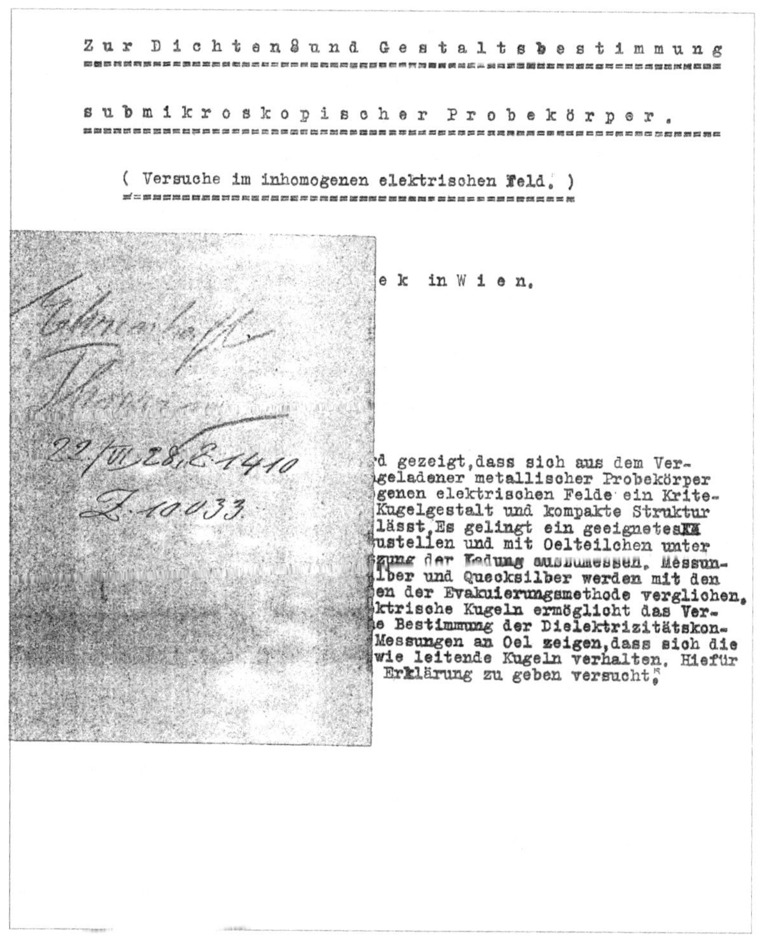

The front page of Placzek's PhD Thesis defended on June 22, 1928, at the University of Vienna. Courtesy of the Central Library of the University of Vienna.

Placzek was attracted to this problem, first in Rome and later in Copenhagen, where at Bohr's suggestion he and Otto Frisch studied the capture of slow neutrons.

Placzek's work in Copenhagen made him a leading authority on neutron scattering and absorption in matter. In a series of experiments, Placzek and Frisch discovered that the absorption of neutrons depends strongly on the atomic mass of the material and the velocity of the neutrons, while for slow neutrons and light elements the neutron-capture cross-section is inversely proportional to the velocity. He also worked with Hans Bethe on a theory of neutron absorption resonances, deriving important laws and selection rules, and publishing a seminal paper on resonance reactions in 1937. Papers published later by Placzek with Bohr and Rudolf Peierls dealt with the general theory of nuclear reactions and rank among the classics. For example, the well-known optical theorem, connecting the imaginary part of a scattering amplitude with the total cross-section, bears the names of Bohr, Peierls and Placzek. Using the optical theorem and Bohr's drop model of the nucleus, the trio developed a fundamental theory of neutron-induced nuclear reactions.

Hitler's preparations for a systematic occupation of all countries bordering Germany endangered some members of Bohr's international team, including Placzek. The Anschluss of Austria in the spring of 1938 and the seizure of a large region from Czechoslovakia through the Munich treaty in September of that year left no-one in the dark. Bohr decided to move part of his Copenhagen Institute to the other side of the Atlantic. Placzek left Copenhagen for the US in January 1939 and in Princeton, at the beginning of February, he met Bohr, who had been waiting for him impatiently.

Across the Atlantic

Placzek's first encounter with Bohr on the other side of the Atlantic provides an interesting illustration of their personal relationship and scientific collaboration. While sources (e.g. Moore 1966, Wheeler and Ford 2000) do not agree exactly on what was discussed during breakfast on February 3, 1939, they do seem to have the same opinion on the following. Bohr found Placzek – "the institute's always stimulating Bohemian" – sitting with Léon Rosenfeld. Their discussion focused on some exciting news from Europe. First, Frisch and Lise Meitner had recently suggested that most of the transuranic elements were produced by a new type of nuclear reaction – the capture of neutrons from uranium fission. Second, Placzek had suggested to Frisch in Copenhagen how he might confirm the existence of fission in a straightforward way, which Frisch promptly did on 13 January 1939.

Figure 5.1 George Placzek, J. Blaton, and Rudolf Peierls, Copenhagen, 1947. Credit: Niels Bohr Archive, Copenhagen.

Bohr, listening attentively, looked up with a big smile: "For one good thing, we're free of transuranic elements." Placzek, the sceptic, 20 years younger than Bohr, commented: "Yes, but now you're in a worse mess. How can you reconcile it with your view of nuclear reactions?" How, he asked, was Bohr going to explain why slow rather than fast neutrons should cause uranium to fission? Why should slow neutrons induce a modest fissioning in uranium, but be captured in thorium?

Bohr suddenly went pale, took Rosenfeld and set off across the campus to his office. He went to the blackboard and worked rapidly, making some rough sketches. In about ten minutes he stopped; he had the answer to the problem posed by Placzek, related to the fissioning of the nuclei. The fission cross-section for slow neutrons must be due to the small amount of the isotope uranium-235, the cross-section increasing as the wavelength of the neutrons increases with decreasing energy.

From 1939 to 1942 Placzek was professor at Cornell University, Ithaca. Later he went to Montreal and Los Alamos, where he contributed to solving problems related to the moderation of neutrons in matter. He was apparently the only citizen of Czechoslovakia to take part in the Manhattan Project, being head of the Theory Group in Chalk River near Montreal, and then in Los Alamos since May 1945. Later he worked for some time in the General Electric Company in Schenectady, and in 1948, he obtained a permanent position at the Institute for Advanced Study in Princeton.

In the last years of his life, Placzek went into more depth with his analysis of the elastic and inelastic scattering of light particles in liquids and crystals, aimed at investigating the physical properties of these media. Albert Messiah and Léon Van Hove were among his collaborators and friends in this period (Van Hove, 1956, Messiah, 1991). During this time, he also visited Europe to lecture on the moderation and diffusion of neutrons and in 1954, the book *Introduction to the Theory of Neutron Diffusion*, by K. M. Case, Frederic de Hoffmann and Placzek was published, based on a lecture course Placzek gave in Santa Monica and Los Angeles in the summer of 1949. In autumn 1955, Placzek died in Zürich while he was planning a several-month lecture tour through Italy for the academic year 1955-56 funded by the Guggenheim Foundation.

During his scientific career, Placzek provided a quantum formulation of Raman light scattering, developed the ideas of molecular symmetry and its application in physics, and in collaboration with Bethe, Bohr and Peierls, contributed to the general theory of nuclear reactions. He systematically studied the behavior of neutrons in nuclear reactions, neutron resonances, scattering and diffusion in matter, and the moderation and absorption of neutrons in crystals and liquids. He was among the first to suggest, independently of several others, that graphite might be used to moderate neutrons. Yet his name is not well known.

Figure 5.2   The Guggenheim file.

The importance of publication

Those who knew Placzek personally agree that his discoveries and results in physics, rich and important as they were, were not sufficiently published; only a small portion of his results appeared in print. As Van Hove points out, Landau's and Placzek's classic results on the scattering of a monochromatic wave in liquids were published in an incomplete form, and were only later discussed in more detail by Yakov Frenkel in his Kinetic Theory of Liquids. The same holds for the results on the general theory of nuclear reactions obtained by Bohr, Peierls and Placzek. Amaldi wrote about Placzek in 1956: "The redaction of an article represented an immense effort for him; even important results of his, which he had formulated clearly and in definite form, often remained unpublished."

This trait of Placzek's corresponded to his desire always to deepen his analyses of the phenomena he studied. Moreover, many of his original ideas are implicitly contained in other papers. As an unmerciful critic, Placzek often served as the scientific conscience for his colleagues, stimulating their invention, criticizing their work and forcing them to formulate their scientific ideas clearly. He had a number of characteristics that made him a welcome collaborator and team member: a highly developed critical sense, an ability to understand new problems quickly and to confront them with relevant facts, an unselfish willingness to offer advice and, last but not least, a generous disinterest in participating in the result.

These attributes, as Amaldi explains, reflected Placzek's character. He excelled in general erudition and in a culture anchored in fundamental ideas. He easily learned foreign languages and felt great affection for small nations. He was noted for strict moral principles and a great sense of tolerance.

What little remains

Despite Placzek's long collaboration with Bohr there are only a few photographs and letters in the Niels Bohr Archive in Copenhagen. One possible explanation is that, before moving to America, Bohr obliterated all traces that could help the Nazis pursue the families of those who had fled from occupied countries. In Placzek's fatherland, by the end of 2004, only a few documents had been found: a note in the register of births (containing the names of his parents and grandparents, the godfathers, the rabbi and the midwife), the regular notes of his studies in the *Staatsgymnasium* high school, and the distasteful entries in the book of the right of domicile. Placzek's parents and his sister died in concentration camps during the Second World War, while his brother died eight days after the Nazis' invasion of Czechoslovakia in March 1939. Recently, however, in connection with the double anniversary of Placzek's birth and death, many interesting documents have been found about his relatives and the history of the whole family in Moravia and North America (Gottvald, 2005). These will be presented at the symposium.

The author is greatly indebted to Ugo Amaldi, Giuseppe Cocconi, Torleif Ericson, Aleš Gottvald, André Martin, Michelle Mazerand, Jack Steinberger, Valentin Telegdi and Jenny Van Hove for providing valuable information relating to George Placzek.

Further reading

For information about the Symposium in Memory of George Placzek (1905-1955) see http://dumbell.physics.muni.cz/placzek/.

References

E. Amaldi, 1956, *La Ricerca Scientifica*, 26, 2037.

J. Fischer, 1985, *Cs. cas. fyz. A 35, 607*, in Czech.

O. R. Frisch, 1979, *What Little I Remember*, (Cambridge University Press).

A. Gottvald, 2005, *Cs. cas. fyz.*, 55, 275, in Czech.

A. Messiah, 1991. *The early post-war period of Léon Van Hove, A recollection Scientific Highlights in Memory of Léon Van Hove*, Ed. F. Nicodemi, (World Scientific, Singapore).

R. Moore, 1966, *Niels Bohr: The Man, his Science and the World They Changed*, (Alfred A Knopf).

L. Van Hove, 1956, *Nuclear Physics*, 1, 623.

Georg Placzek on Radio Czech[4]
Christian Falvey interviews Jan Fischer

He died young, at the age of 50, and for a long time hardly anyone in Czechoslovakia knew of his importance. Some 60 years after he left his homeland forever, he was brought to public attention by Czech physicist Jan Fischer, who among other things was impressed to learn there had been much grief among the world's great physicists when Placzek died in 1955.

Georg Placzek

"He was appreciated much more abroad than in his home country – in his home country he was completely unknown. So I became proud of my countryman and I thought this was very nice that people abroad appreciated him and this made me interested in him."

It's a strong testament to half a century's seclusion that the Placzeks' name had been forgotten in their hometown of Brno. Georg's great-grandfather and grandfather had been important Moravian rabbis, the latter being of a scientific bent, a close friend of Gregor Mendel's and a pen-pal of Charles Darwin's. His father ran a large textile mill, and was honored for the humanity he had shown the community in the First War through the care and opportunities he afforded the poor. It was a popular family that was well-knit into its multinational society.

"Placzek grew up in a milieu of mixed nations: Czechs, Moravians, Jews and Germans. And I think it is a unique milieu, where Gregor Mendel grew up, or Sigmund Freud, the philosopher Edmund Husserl, the composer Gustav Mahler, and Franz Kafka – of course in Prague. So I think that these mixed nations are very fruitful ground for raising interesting people."

Placzek's earliest penchant was for language, and he spoke around ten fluently. He was well-conversant in matters of culture throughout his life, and he was a jovial and sociable person. When he completed his studies in physics in Prague and Vienna with resounding success, his intellectual versatility found him a warm welcome among some of the greatest scientists in history: Werner Heisenberg, Enrico Fermi, Otto Frisch, Robert Oppenheimer, Edward Teller, and of course, the father of

---

[4]January 20, 2010, 17:28. http://www.radio.cz/en/section/czechs/georg-placzek-a-critical-link-in-the-making-of-the-atom-bomb

quantum mechanics, Niels Bohr, in whose Copenhagen research center Placzek spent six years.

Niels Bohr

"He called Placzek the 'ever-stimulating Bohemian'. Because they were two different types of people, I think, the best combination for theoretical physics: one who has new ideas – maybe crazy ideas – and on the other hand someone like Placzek, who was very critical. And such a combination of authors is excellent. Placzek, through his criticism, was able to stimulate Bohr to obtain new results that would not have been obtained without it."

He also had a critical mind for politics. First a radical leftist and later a staunch critic of the radical left, Placzek's political views were mutable but always strong, it seems. In 1934, he insisted to his friends and colleagues that Hitler would start a war within five years. He tried to persuade his family to leave, but they didn't believe him until it was too late.

"During the second war, all of them died. They were simply exterminated, his parents. His brother committed suicide after trying to emigrate to England in '39. He asked his people from the cash office to give him money to fly to London, and they – suddenly – turned out to be Nazis; they had been his friends but now they were Nazis. So they refused to give him his own money. And he went into the next room and he shot himself."

Placzek's parents were apparently killed in Auschwitz; his sister ended up in a concentration camp and was never heard from again. After paying a visit to Moravia in a final attempt to get them to emigrate with him, Placzek moved to America along with much of Niels Bohr's Copenhagen group.

"Bohr decided to move the Jewish part of his school, his group, to the United States. Niels Bohr was a powerful professor, they were all young – 25 to 30 years old – Niels Bohr was already in his mid-50s, or something like that, so he was able to move these endangered personalities to the United States."

In America, first at Princeton, Placzek continued his work with the greatest minds of the golden age of physics and they were strongly influenced by him, professionally and intellectually. He was said to be single-handedly responsible for converting many a colleague from their communist streaks, such as Robert Oppenheimer (though seeing the light failed to save him from the Red Scare inquisitions). His great talent it seems was in noticing the details that other people, even the sharpest minds on the planet, missed.

"His main benefit was his criticism of other people's works. They wrote papers and he would read them and say 'no, no, no, not this, not this,' he was very critical, and he was of great help to many people just in finding their errors, or their misconceptions or bad formulations. So he was sought after, looked to, for his criticism, and the papers that passed through his filter were of high quality."

Georg Placzek's work was primarily on light and its dispersion, but his greatest impact regarded the neutron. He was taken on board to work on the Manhattan Project, the race to build an atomic bomb before the Germans, because he had the highest credentials. Niels Bohr credited him with clarifying how the atom is split, Otto Frisch credited him with giving the fundamental idea for proving nuclear fission – all of which added up to one of the most important moments in human history.

The geniuses assembled in the desert in goggles and sunscreen on the morning of the Trinity test recorded the tremendous emotions of the moment as the countdown was read out (Tchaikovsky's Nutcracker Suite crackling in the background due to radio interference from a storm) and they first witnessed the awful power of their creation.

As the head of the Manhattan Project's theoretical department, Placzek was there among the 260 observers, but there is no record of his thoughts or impressions. He tended to stay in the shadow of things, seemingly having little need for public recognition. Almost unthinkable for a scientist, he published little and was offered and spurned co-authorships of even important theories he had been key to refining.

"And so there is a suspicion that I have found also in literature that the main contribution to many papers of famous people was thanks to Placzek's criticism. Many of his results remained unpublished. He was able to calculate, to obtain the interesting results, but then to sell it – to write an article – was extremely difficult for him. He would start writing, then not be satisfied and start again..."

This is why an important scientist remained unsung for decades after his death the circumstances of which, incidentally, remain as furtive as the details of much of his life. In the early 1950s Placzek began spending more time in Europe and in 1955 he planned to spend the academic year in Rome. In October of that year, he left for a weekend and gave no address, and was never seen alive again.

"He died suddenly, nobody knows why. He always seemed to be a very happy person, but there is concern that he committed suicide – no one knows, but that's what I was told by people who knew him very well. Maybe suicide was inherited in his family. But I was told he had high blood pressure, and the only medicine available for that condition in the 1950s was something that caused depressive side effects."

Georg Placzek has only come to be a well-known name in recent years, and the years to come may add to that renown, as time uncovers more about the contributions this great scientist made to physics and to history.

The Early Post-War Period of Léon Van Hove.
A Recollection [5]

A. Messiah

I first met Léon van Hove in the fall of 1949, as we both were visitors for a year at the Institute for Advanced Study, Princeton. We became very close friends. In this talk, I would like first to recall our year in Princeton, portray Léon as he appeared to me at this time of our youth, and tell about the collaboration that he started with George Placzek on neutron physics, a collaboration which lasted some five years and proved unusually fruitful. Then, I will account for his contributions in the field. It culminated with the establishment of the formalism of the so-called pair-distribution function in space and time, which is now of current use in this field of physics. Finally, I will conclude by recalling the impact of his achievements of this period on the community of physicists, and more specifically, through personnel contacts, on our community in France.

***

As I said, I met Léon for the first time when I arrived at Princeton on September 1949. We were both married and fathers of a baby. Like many visiting members with family, we were housed in barracks erected on the grounds of the Institute. Thus we were neighbors, both young men in our twenties – he, in fact, was three years younger – with many common interests, furthermore both french-speaking, a precious advantage for me at a time when my practice of broken English was rather weak. We rapidly became close friends, all the more so since he was a happy character, always ready to talk and to listen, and interested in a wide variety of subjects. At this time, the Institute for Advanced Study was an extraordinary meeting ground of outstanding personalities.

Oppenheimer was the Director, assisted by Abraham Pais. Among permanent members were J. von Neumann (devoted then to elaborating computers), A. Einstein (who could be seen from time to time wandering meditatively across the parks). Associated with the School of Physics were E. P. Wigner, professor at Princeton University, and distinguished visitors for a term among which was W. Pauli. The great

[5]Published in *Twentieth Century Physics: Essays and Recollections*, a selection of historical writings – Edoardo Amaldi Foundation series, vol. 3. Eds. Edoardo Amaldi, Giovanni Battimelli, and Giovanni Paoloni, (World Scientific, Singapore, 1998).

Niels Bohr stayed for a whole month. People like H. A. Bethe, R. P. Feynman came to give seminars. Visiting members for the year were a bunch of young men highly-selected [6] and bright, coming either from the U.S. (Schwinger's students from Harvard, or people like C.N. Yang who had started his American curriculum as an experimentalist and then worked under Fermi at Chicago) or from abroad like Res Jost or others. All these people formed an extraordinary lively community.

George Placzek, one of the permanent members of the School of Physics, deserves a particular mention. A typical member of the *intelligentsia* of Central Europe, he had participated in the great venture of physics since the establishment of Quantum Theory, along with people of his generation such as Bethe, Fermi, Peierls, Weisskopf, Wigner, etc. At the Institute, he represented the field of theoretical research in closest contact with experimental activities. At this time, he was mostly interested in the possibility of using neutrons as a probe for studying complex materials. He was looking for assistants and collaborators. He took me and Léon, together with B. Nijboer, a Dutch visiting member, to work with him. The problem that he proposed to me was rather simple, whereas Nijboer and Léon were associated in much more advanced researches.

I stayed very close to Léon Van Hove during this year in Princeton, although we never worked specifically on the same problems. Let me portray him, as he looked to me at this time. From the very beginning, I saw him as a scientist of great stature, and I considered as a great chance to get a friend of such prominent qualities. Although somewhat younger, he was for me an elder brother, a very fine elder brother that some good fortune had put into my life.

Talking with him was pleasant and rewarding. He used to speak rather slowly, carefully choosing the right word, the good way to make himself understood, proceeding step by step, like a mountain climber who progresses methodically and safely up to the summit. The numerous conversations that we had were for me a source of continuous enrichment.

The extent of his scientific culture was impressive. It ranged from pure mathematics to the most recent advances in theoretical physics.

---

[6]Why I, a novice in physics, was accepted there, is still a mystery for me. I had volunteered in the Free French Forces during the war, and resumed academic studies after an interruption of five years. Léon, three years younger, had a different fate. He lived in Brussels through the whole tragic period of the Nazi occupation. The Université Libre de Bruxelles, in which he had entered, was closed at that time by the Nazis, but continued to work clandestinely without interruption.

He had got his very first training as a mathematician, and shifted later on to theoretical physics. He was a fine, well-gifted mathematician. His broad culture in many fields of mathematics was instrumental in getting him quickly acquainted with the fundamentals of modern physics. On the one hand, he was familiar with statistical mechanics, including transport phenomena and speculations on thermodynamics of irreversible processes. But he had also a profound understanding of Quantum Theory. He knew about the most recent advances like those in QED following the works of Feynman, Schwinger and Tomonoga, which substantiated the concept of renormalization and permitted to account for the recently measured higher-order corrections. Besides he was well aware of the epistemological questions raised by the fundamental principles of Quantum Theory, and knew the tenets of the ever-going dispute about them.

His interest in the whole field of physics was unfailing. Bright as he was, he had not much difficulty in keeping himself well informed of the most significant progresses. Princeton, at this time, was an ideal place, with its gathering of so many prominent people. Leon did not miss the opportunity. Georges Placzek, whom he met regularly for his current work on neutrons, was himself a man of learning, and Leon, no doubt, greatly benefited of his long experience. But Leon met also many others. And, since he loved to communicate, he often spoke with friends, at his familiar slow pace and always with great clarity, of whom he had met and what he had learned.

One day, he told me that he had arranged to meet Einstein. This was a long visit, from which he came back deeply impressed. Contrary to expectations, Einstein appeared to him as perfectly informed of all the progresses of Quantum Physics including the latest. And Léon got from him, first-hand, his deeply rooted conviction that Quantum Physics was extraordinarily useful, but nevertheless incomplete theory. Léon told me of many other contacts. Let me mention those related to a purely mathematical investigation, the idea of which came to him when he was in Princeton. It concerned the relation between the group of canonical transformations of Classical Mechanics and the group of unitary transformations of Quantum Mechanics. This looked to him a good subject for a "Thèse d'Agrégation" to be presented at the Université Libre de Bruxelles (which he actually did in April 1951, see [1]). He pursued this research program of his own up to completion during the year, and it gave him the opportunity of fruitful contacts with such mathematicians as J. von Neumann and V. Bargmann, and with such prominent physicists as O. Klein and W. Pauli.

Léon's "Memoire de Thèse d'Aggregation" is a contribution of high-level mathematics. It requires a mastery in group theory and in the handling of abstract mathematical concepts. Conversely, the collaboration with Placzek drove him "down-to-earth," far away from abstractions, toward the elucidation of specific physical phenomena. A completely different game. It demands a knowledge of order of magnitudes, an ability to guess the good approximations and to find the suitable mathematical framework to represent phenomena. That Leon was able to play with equal success both types of game is characteristic of his personality.

His collaboration with Georges Placzek on neutron physics lasted several years and culminated with one of his major contributions to physics, namely the introduction of the so-called pair-correlation function in space and time, $G(r,t)$, in the analysis of neutron scattering data. I will now recall this "love story" of Léon van Hove with neutron physics.

***

Léon's involvement in the field started at a time, when thermal neutron reactors had not yet started their prodigious proliferation all over the world. But the newly available neutron beams looked already to be a promising tool to probe dense matter. Teams of experimentalists had started pioneering work in this direction. Georges Placzek devoted himself fully to all the theoretical aspects of this field of research.

References

[1] L. van Hove, "Sur certaines représentations unitairee d'un groupe infini de transformations", Thèse présentée le 20 Avril 1951 à la Faculté des Sciences de l'Université libre de Bruxelles pour le grade d'agrégé de l'Enseignement Supérieur. Academie Royale de Belgique. Mémoires Tome XXVI Fasc. 6 (1951).

[2] G. Placzek, B.R.A. Nijboer and L. van Hove, "Effects of short wavelength interference on neutron scattering by dense systems of heavy nuclei," *Phys. Rev.*, 82, 392-403 (1951).

[3] B.R.A. Nijboer and L. van Hove, "Radial distribution function of a gas of hard spheres and the superposition approximation," *Phys. Rev.*, 85, 777-83, (1952).

[4] L. van Hove, "The occurrence of singularities in the elastic frequency distribution of a crystal," *Phys. Rev.*, 89, 1189-93 (1953).

[5] G. Placzek and L. van Hove, "Crystal dynamics and inelastic scattering of neutrons," *Phys. Rev.* 93, 1207-14 (1954).

Excerpts from
The Accused, by Alexander Weissberg (pages 82, 114-118)[7]

Viktor Weisskopf had come to Moscow. Niels Bohr had asked him and the physicist Georg Placzek to let him have a report on Soviet science, and to find out whether it would be possible to send many prominent German physicists to the Soviet Union, men who were now unable to work in Germany because they were anti-Nazis.

February 1937

[...]

"Do you think you can lead us by the nose much longer? We've got you where we want you, I tell you. Who sent you here?"

I still remained silent and he sprang up and almost danced with rage.

"Who did your dirty work for you here?" he demanded. "What instructions did the Trotskyist agent Placzek bring you from abroad?"

So Placzek was a Trotskyist agent, too! I still said nothing. There was nothing to say. He calmed down after a while and went on:

"Accused Weissberg, if you continue to refuse to answer questions you will be guilty of an anti-Soviet action right here in the office of the examiner. You would have to pay for that dearly. We have very effective ways and means of dealing with people who sabotage our inquiries."

"Citizen Examiner, I have not the slightest intention of sabotaging your inquiries; on the contrary, I want to facilitate them. But you aren't putting any definite questions to me to which I could give you definite replies."

"I have asked you what counterrevolutionary instructions were given to you by the agents of international Trotskyism. Did Placzek bring you the program of their new espionage organization, the Fourth International?"

"As far as I know Placzek has nothing whatever to do with the Fourth International."

"How much longer are you going to say that black's white? How much longer are you going to lie to us in this shameless fashion? Do you want to deny the obvious facts recognized by everyone? Placzek

---

[7]Alexander Weissberg-Cybulski, *Hexensabbat*, (Frankfurt, Frankfurter Hefte, 1951, in German), English translation: Alexander Weissberg, *The Accused*, (New York, Simon and Schuster, 1951) and *Conspiracy of Silence*, (Hamish Hamilton, London, 1952) A modern German edition was released recently in Austria under the title *Im Verhör: Ein Überlebender der Stalinistischen Sauberungen Berichtet*, (Europa Verlag, 1993). It is instructive to compare Weissberg's recollections with Konrad Weisselberg's file, see Chapter 2.

insolently supported the Fourth International in the Institute. Do you dare deny that? We have four witnesses."

"Placzek isn't really political at all. He was here for a few days and he talked a lot of airy nonsense, that's all. He's a first-class physicist, and therefore he was invited to pay us a visit. He'd be about as useful as a secret agent as I would be a tightrope walker."

"We have examined the Placzek case from all angles," he declared weightily. "It's one of your favorite tricks to pretend that your agents are all simple persons who couldn't say boo to a goose. But you can't fool us. We expose the counterrevolutionary no matter what the mask."

The case of Placzek was a striking example of how dangerous a little humor can be in a country of totalitarian dictatorship. Theoretical physicists regarded him as an experimental physicist, while the experimental physicists regarded him as a theoretical physicist. In reality he was something between the two, a sort of liaison officer. *He came from a wealthy family and he didn't have to earn his living, so instead of taking a permanent appointment he went from one famous physicist to the other.* For a few weeks he would stay with Niels Bohr in Copenhagen, and after that perhaps a couple of months with Fermi in Rome. At the same time he was a very amusing fellow and people liked him. He would make the most ridiculous jokes and tell the most absurd stories. We tried to persuade him that he was wasting his talents flitting around the world, and on one occasion he actually agreed to accept a professorship at the Hebrew University in Jerusalem, but in his inaugural lecture to a packed hall he thought it necessary to inform his audience that the Jews weren't really a people at all or Hebrew a language, and that it was only by a miracle of God – such as one would, of course, expect in the Holy Land – that the children on the streets could make themselves understood to each other. One can imagine what sort of effect that kind of airy persiflage had on the newly baked nationalism of the Jews. Placzek stood it for a year, and then he packed up his things and began his wanderings again. He went back to Germany, where he made the most outrageous remarks about Hitler, who was already Reichschancellor. *Lèse-majesté* of that sort was dangerous to life and limb, and Placzek's friends breathed again when he moved on. He came to visit us in Kharkov on a number of occasions. *The last time I met him was at the beginning of December, 1936, in Moscow.* I was living in the Hotel Moskva at the time and he rang me up to arrange a meeting.

"Where shall we meet?" I asked.

"In the street named after the traitor," he suggested. I knew what he meant, the Bolshaya Dimitrovka. That was its name before the revolution and it had nothing whatever to do with Georgi Dimitrov, who, although not a German, had been charged with treason at the Leipzig trial, and was now Chairman of the Communist International. Placzek was only joking; in reality he had a great admiration for Dimitrov on account of his heroic attitude at the trial – but what about my telephone censor? I slammed down the receiver angrily; the man was impossible. When we met I reproached him for his lack of consideration. He looked at me pityingly.

"This country's going to the dogs," he declared. "All sense of humor is disappearing. I can see the decline in you. Twelve months ago you were almost human." It was quite impossible to make him see reason or to curb his caustic tongue. I gave up trying. I stayed in Moscow and he went off to Kharkov with my permission to use my rooms. When I came back I learned what had happened.

In a discussion with Barbara and Martin Ruhemann, Placzek had declared that the Third International was a waste of time now that Lenin wasn't there to keep it in order. All revolutionary forces ought now to be organized in Trotsky's Fourth International. Barbara Ruhemann was a convinced Stalinist and she opposed the idea vehemently. Placzek was delighted to have drawn her and he provoked her more and more. Martin Ruhemann listened quietly and grinned. He knew it was silly to take Placzek seriously. Not so Barbara, who was boiling with indignation, and the next day she told Party Secretary Komarov[8] the whole story. That put Komarov in a cleft stick. He had no interest in baiting Trotskyists, and he was very sorry Barbara had told him. However, if she had told him she might tell others, and if he suppressed the report the GPU might get to hear of it and pounce on him. That could cost him his post, his Party membership and perhaps his liberty. Unwillingly, therefore, he sat down and wrote out his report.

The GPU did nothing about Placzek. He was a foreigner on a short lecture tour and in a few weeks he would be gone, but the stick was good enough to beat me with. For some years they had been collecting material against me without coming across anything of any importance,

---

[8] Pyotr Nikolaevich Komarov, a UPTI graduate student, worked as a member of the OSGO group. He was arrested by the NKVD in 1937 and perished in Gulag. OSGO is the Russian abbreviation for *Experimental Station for Deep Freezing*, a cryogenic facility which was constructed at UPTI under the directorship of A. Weissberg.

and now here was a real live Trotskyist, obviously an agent of the Fourth International, a pet subject of abuse in the Soviet press since the days of the Zinoviev trial, and the man was a close friend of mine and I had placed my rooms at his disposal.

When I came back and heard the story, I realized the danger at once, and I took Placzek to one side.

"Listen," I said, "I'm not reproaching you. You just don't understand the situation here. However, it's gone so far this time that I've got to disassociate myself from you for my own safety. Don't misunderstand my motives therefore when I must now ask you to go and live somewhere else."

After that I went to the Institute and asked them to find Placzek a room in a hotel, and he moved out the very same day. As far as I was concerned that settled the matter, but not for the GPU. At last it had an objective fact it could use against me. Most of the indictments were based on statements extorted from arrested men. Conversations were invented which had never taken place; facts were invoked which had nothing to do with the truth. But the discussion between Placzek and Barbara Ruhemann had really taken place. It was too good to be missed.

It was true that I had not taken part in it or even been present at it. In addition, the things Placzek had said were not punishable by Soviet law. He had expressed political opinions which were not in accordance with those of the Party leadership, but he was not even a Communist. There was no room for any charge of treason. And where did I come in anyhow? I had what was a big flat by Soviet standards, and therefore the Institute expected me to put up foreign visitors from time to time. Was I therefore responsible for their political opinions?

For my examiner the situation was perfectly clear. Placzek was an agent of the Fourth International. The Fourth International was an organization which sent spies and saboteurs into the Soviet Union to work in the interests of capitalism. Placzek had stayed in my flat at my invitation. Therefore I was a spy and a saboteur. What could be clearer? All that remained was to get me to confess that I was, in fact, a spy and a saboteur, and the preliminary investigation could be satisfactorily concluded. I tried to convince my examiner that Placzek was a confirmed joker and that it was absurd for the Soviet authorities to take him seriously.

"You must try to understand," I explained patiently, "that Placzek was brought up in countries where the sense of political responsibility as it is understood here just does not exist. When he is talking he says whatever comes into his mind and sounds amusing. The next time he may very well say the opposite. He never has to think of his position or his liberty no matter how he talks."

"Placzek is an inveterate Trotskyist," answered my examiner inexorably. "We have exact material to show it. He came to give you secret instructions. Now, what instructions did he give you?"

I gave up. Either he couldn't or wouldn't understand. Whichever it was, the situation was equally hopeless. He shouted, stormed and threatened. He demanded that I should confess that Placzek had given me secret counterrevolutionary instructions, and he also wanted to know to whom I had handed them on, and how much had already been carried out. The interrogation became more and more monotonous. He kept repeating the same words and I kept answering in the same words. Finally I was too exhausted to say anything, and then he called off the interrogation and sent me back to my cell.

I undressed and went to bed. It was about eleven o'clock. I had been in bed for perhaps half an hour when a warden opened the wicket.

"Your name?" he asked.

He had the usual chit in his hand for the examination of a prisoner. I thought there must be some mistake; I had just had a six-hour interrogation. I told him my name. "Get dressed," he said. "Interrogation."

I got dressed and followed him. I was still a little dazed when I sat down opposite the examiner.

"Are you prepared to confess at last or do you intend to continue the struggle against us?"

It was too much, and I said nothing.

"You fascist swine," he shouted. "How much longer do you think you can play this game with us?"

I still said nothing.

"Take care. In our cellars you'll lose your dumb insolence quickly enough, I warn you. Now come on, out with it."

I remained silent.

"Where is the illegal material Placzek gave you? Don't try to treat us as though we were fools. A Trotskyist spy comes to you and all you talk about is the weather, what?"

I could still find nothing to say.

"Alexander Semyonovitch Weissberg, we have treated you humanely up to now. We wanted to make things easy for you. But you mustn't think we haven't ways and means of dealing with refractory prisoners. We have all the material we want to hand you over to the courts. Your discussions with Placzek, Anders and Komarov are all known to us."

"With Komarov!" I exclaimed at last. "What's the matter with Komarov? He was our Party Secretary."

"We'll talk about that later. All you've got to do now is to sign this."

He had drawn up the deposition about the Placzek affair in such a way that every detail sounded suspicious. There were no direct falsehoods, but the whole thing had such a slant that anyone reading it would have thought that a plot against the state had been revealed. I can no longer recall the details and it was a method adopted only in the first, and more humane, phase of my imprisonment [...]

Excerpt from

Recollections of Lev Davidovich Landau, by A. I. Akhiezer [9]

Shortly before this, the government had decided to reinstate the awarding of scientific degrees and ranks, which had been stopped after the revolution. At our institute Landau, Obreimov, Leipunsky and Sinelnikov were awarded the degree of doctor of sciences without having to defend dissertations. Soon thereafter Lifshitz defended his dissertation, and then, in 1936, Tisza and I defended ours. These were the first three dissertations to come out of Landau's school.

At Tisza's and my defense, the appointed reviewers were Igor Tamm from Moscow and George Placzek, another German emigré then visiting at our institute. After the defense, there was a wonderful dinner at Landau's apartment. It seemed to us a serene time, even though it was already two years after the vicious murder of Sergei Kirov, secretary of the Leningrad Regional Party Committee. But none of us at the time understood the meaning of that murder. (It eventually became clear that Kirov was an early victim of Stalin's terror.)

Later in 1936, Landau, Pomeranchuk and I were listening to Stalin on the radio, announcing his new constitution. After that historic speech, Landau commented that this marked the beginning of "a new, good era." At that time Landau was "red," and he didn't tolerate the expression of any anti-Soviet opinion.

---

[9] Alexander I. Akhiezer, *Physics Today*, 47(6), 35 (1994).

Excerpts from

What Little I Remember, by Otto Frisch, pages 81-83[10]

Denmark 1934-1939

Niels Bohr's institute was international in quite a different way. It was the Mecca of the world's theoretical physicists. There were many foreigners about, but they kept changing; most of them were transient visitors who would come and give a seminar, a talk or two, and disappear again. One of the first talks I heard was given by George Gamow. I cautiously enquired what language the famous Russian physicist was going to speak and was told 'Danish; but don't worry, you'll understand him.' How could I, having been in Denmark only a few days? I hadn't even started taking Danish lessons. But all the same I understood Gamow; he peppered his Danish with English and German words, gesticulated and made funny drawings. He really knew how to communicate and was very entertaining, as anybody will understand who knows his Mr. Tompkins' books in which the mysteries of physics are expounded to laymen in a very amusing if not always quite accurate way.

Another person with whom I became closely acquainted was George Placzek, a Bohemian in every sense of the word. He came from what later became known as Czechoslovakia, had studied in Vienna and been all over Europe. I reckon that by the time I left Copenhagen he spoke ten languages more or less fluently and with a fine range of naughty verse in most of them. When I met him he had just accepted an appointment as Professor of Theoretical Physics at the new University of Jerusalem and was due to leave shortly.

To pack his belongings was a tiresome chore, and he asked me to keep him company to prevent him from falling asleep. He kept up a barrage of talk and comment while I sat there and idly – thumbed through the books he hadn't packed yet. At one point he gave a sudden shout and held up a piece of paper: the receipt for a sum of 100 German marks which he had left as a deposit with the Berlin University Library, a deposit which he had quite forgotten but would now be able to recover on passing through Berlin. That happy windfall had to be celebrated; toothbrush glasses were found, and brandy was drunk from them. After the ceremony, the piece of paper had of course disappeared and Placzek

---

[10]Otto Frisch, *What Little I Remember*, (Cambridge University Press, 1979).

started to unpack again with many curses and imprecations. Finally he found it again and held it under my nose so that if it disappeared once more at least there should be a witness that the whole thing hadn't been a dream.

The packing went on for a couple of days, with slowly increasing frenzy as his departure drew near. On the last evening we were both trying to stuff the rest of his belongings into a trunk, and I vividly remember a large eiderdown which kept extruding one corner whenever the other three had been squeezed into the trunk.

In the middle of this he suddenly left me to it and started dictating a letter in Danish to one of his friends who had turned up to help. That letter contained detailed instructions to the carriers, what to do with his various belongings, which things to ship to Israel and which to put in store, either to be forwarded later or to be kept in case he returned to Denmark.

Figure 5.3   Sketch of Placzek by Otto Frisch.

Ten minutes before the train was due to leave we all raced down the stairs with trunk and suitcases and crammed into a taxi, telling the driver to go hellishly fast to the station. During that drive Placzek implored Divinity to let him catch the train, with promises that he would be good in future and never arrive at the railway station less than ten minutes before the train was due. We arrived one minute after the scheduled departure; Placzek ran behind a panting porter who had grabbed his cases. Hopeless it seemed; but the train was still there! The sleeping-car attendant had noticed that one of his flock was missing and had held the train for two minutes; so Placzek caught it.

We later had numerous communications from Israel; how at first he had to fend off the persistent requests from the university authorities that he should give his lectures in Hebrew, which he had not yet learned. They gave him a year to learn Hebrew. (He learned Arabic as well.) When at the end of that year he still wouldn't give lectures in Hebrew – he felt the language couldn't cope with modern physics – and they insisted on it, there was a telegram 'through with jews for ever,' and he came back to Denmark. The scientific language universally used in the laboratory was English; the Danes, by and large, understood it, and certainly the scientists did. All the same it was clear that I would have to learn Danish, and I found a teacher, a charming old lady of eighty who taught me three times a week, with great patience and clarity. Within a few weeks I had acquired a spurious fluency which caused one of my friends to say "Frisch really knows only twenty words of Danish, but he uses them as if he knew a great many more."

***

Excerpts from

Experimental Work with Nuclei, Otto Frisch, pp. 69-72[11]

George Placzek had left for Israel soon after I came to Copenhagen, and then I was the only one with enough knowledge of Italian to give an extempore translation of Fermi's latest results, each time a new issue (or preprint?) of *La Ricerca Scientifica* arrived. Once we had our neutron source, we got rhodium foil and began to repeat and extend Fermi's work; James Franck took an active part for a while. Hevesy noted that quite a thin gold foil became more active on the side where it was struck by slow neutrons and that this effect became more pronounced behind a sheet of cadmium; here was another example of the "neutron groups" reported from Rome.

It must have been late in 1935 that Bohr conceived his idea of a compound nucleus as a long-lived intermediate state in a nuclear reaction. I vividly remember the occasion: Bohr repeatedly (more than usually) interrupted a colloquium speaker who tried to report on a paper (by Hans Bethe, I believe) on the interaction of neutrons with nuclei; then, having got up once more, Bohr sat down again, his face suddenly quite dead. We watched him for several seconds, getting anxious; but then he stood up again and said, with an apologetic smile, "Now I understand it all"; and he outlined the compound nucleus idea.

George Placzek had come back from Israel, preceded by his famous telegram "Through with Jews forever" – they had insisted that he lecture in Hebrew – and asked me to help him with some experiments, making use of a slow neutron detector, a boron-lined ionization chamber, which I had built. The resulting paper was published in *Nature*, in

---

[11]From *Nuclear Physics in Retrospect*, Proc. of the 1977 Symp. on the History of Nuclear Physics in the 1930s, Ed. Roger Stuewer, (University of Minnesota Press, 1979).

the same issue in which Bohr's lecture to the Videnskabernes Selskab of January 27, 1936, was published. I hadn't really understood what Placzek was after until we started to draft our joint account, urged on by Bohr. The aim was to measure the absorption in boron of those slow neutrons which were transmitted by cadmium; it was found to be only a few times smaller than the absorption of the thermal neutrons, and Placzek concluded that the width of the cadmium absorption region was 1 eV or less.

In a paper submitted in November 1935, Leo Szilard had described the selective absorption of neutrons and indeed indicated the basic idea of our experiment in the sentence: "An attempt is now being made to determine the energy value of these absorbing regions by studying the absorption in boron and lithium of the 'highly absorbable' components of the residual beam." I certainly hadn't noticed that (probably not read the paper) and probably Placzek hadn't; he was a bit embarrassed when Szilard wrote him a letter pointing it out, quite gently.

In the summer of 1938, soon after the take-over of Austria, Lise Meitner was informed of an order (by Himmler, I believe) that scientists were not to be allowed to leave Germany. Her application for permission to travel abroad was indeed refused; her dismissal under the racial laws seemed imminent since all her grandparents had been Jews. Peter Debye communicated her plight to his Dutch colleagues, and she was persuaded to travel to Holland without a visa; a Dutch physicist, Coster, met her train at the frontier, having arranged that the Dutch immigration officer would let her in. From Holland (where there was little nuclear physics going on at the time) she went to Copenhagen, where she stayed for a number of days as Niels Bohr's guest; finally she accepted an offer from Manne Siegbahn in Stockholm, where a cyclotron had just been built (the first on the continent of Europe).

I feel embarrassed to tell once again the story how she asked me to join her at a hotel "Kungälv," a few miles north of Gothenburg, where Swedish friends had invited her to Christmas dinner, and how I found her at breakfast poring over a letter from Otto Hahn. In an earlier letter (which seems to have been lost) she had urged him not to publish the alarming result that radium isotopes resulted from neutron bombardment of uranium, at least not until he had checked that result with the greatest care. That check had now been made, with Fritz Strassmann, and had shown that they had isotopes of barium, not radium. As we talked this over, we came to consider Bohr's analogy between a nucleus and a drop of liquid; I estimated the loss of surface tension due to the nuclear charge and the energy freed in the electrostatic repulsion between the two resulting droplets, overestimated because the fragments were assumed to be of equal size; at the same time Lise Meitner estimated the energy freed according to the packing fraction, also overestimated because the energy subsequently lost by several β-decays was included. So we got good agreement, about 200 MeV, indicating that the process was classically possible; the true figure is more like 160 MeV.

When I told Bohr a few days later, back in Copenhagen, he instantly agreed with our interpretation; he had been aware of the reduced surface tension of heavy nuclei and had correlated it with the narrower spacing of their lowest levels; he called himself a fool for not having predicted fission! I proceeded to draft a joint paper and discuss it with Lise Meitner (back in Stockholm) over the telephone. But it took Placzek to make me think about experimental proof. He was skeptical; the idea that uranium should be liable to fission as well as α-decay, he said, was like dissecting a man killed by a fallen brick and finding that he would have soon died from cancer – an unlikely coincidence!

Placzek suggested a cloud chamber to look for those fission fragments; that was done much later by others and revealed a wealth of secondaries, making the tracks look like Christmas trees. But I quickly made up a very simple ionization chamber – two parallel metal plates separated by a glass ring about 1 cm high; a foil with a layer of uranium hydroxide could be placed on the bottom plate. One of our neutron sources, 100 mg radium mixed with beryllium, was placed close to the chamber; the bias of the amplifier was adjusted until both the α-ray pulses and the noise caused by the γ-rays from the source were suppressed.

My notes of January 13-14 (found and inspected recently for the first time after thirty-eight years) showed that measurements were started in the afternoon, and pulses at about the predicted amplitude and frequency (one or two per minute) were seen within a few hours; but the measurements were continued until six in the morning to verify that the apparatus was working consistently, that the counting rate was only a few counts per hour in the absence of either the uranium or the neutron source, and that the response to variations of the bias and the insertion of paraffin were as expected.

On Monday, January 16, I observed fission pulses from thorium but not from lead. On the same day I sent two letters to the editor of Nature, one by Lise Meitner and myself in which our explanation of the results of Hahn and Strassmann was put forward and the term "nuclear fission" proposed; the other in my name only, reporting the results described above. They appeared on February 11 and 18 respectively.

References

[1]. Lise Meitner and O. Frisch, "Disintegration of Uranium by Neutrons. A New Type of Nuclear Reaction," *Nature*, **143**, 239-240 (1939).

[2]. O. Frisch, "Physical Evidence for the Division of Heavy Nuclei under Neutron Bombardment," *Nature*, **143**, 276 (1939).

Excerpts from
The Joy of Insight by V. Weisskopf[12]

Leipzig 1931

Because Heisenberg was not yet married, he spent a lot of time with his students. Among other things, he was a great ping-pong player. He had one Japanese co-worker, Yoshio Nishina, who played better than he and who beat him regularly. I recall one occasion when Heisenberg, who was not a good loser, disappeared for three days after a defeat by Nishina. Heisenberg was also an amateur pianist. I still recall how he played Beethoven's extremely difficult Hammerklavier Sonata for us. His performance was technically perfect but almost completely devoid of passion.

Once again I found a wonderful group of young people gathered around a great teacher. Among these were Felix Bloch and George Placzek, who became my close friends.

With Niels Bohr in Copenhagen

George Placzek, a native of Czechoslovakia who grew up in Vienna, was one of the most extraordinary people I have ever known. We considered him a wise man with whom we were eager to discuss personal, political, and scientific problems. Placzek had an especially clear insight into the problems of the time, whether in physics or politics. But for some reason he was never able to find the right style of life for himself, and his personal life was not a happy one.

Impeccably honest and loyal to his friends, Placzek was known for his intellect and scientific prowess. His somewhat outrageous sense of humor and lifestyle inspired dozens of "Placzek stories." He was a truly unique man, the kind of person who is constantly involved in some sort of antic adventure. For example, after being in Copenhagen for some time, Placzek got a little apartment in the institute. He used to sleep very late, often until noon, when he would get up and go down to get his mail. The sight of Placzek still in his rumpled pajamas at midday in the library offended several of the female students, who complained to Mrs. Bohr. She asked me, as Placzek's friend, to intervene. Placzek assured me that there was never anyone there when he came for his

---

[12]From Victor Weisskopf, *The Joy of Insight*, (Basic Books, 1991), pp. 49, 68-69,101, 114-115, 119, 133-134.

mail. All the other physicists arrived later, he insisted. I said that there were always several students in the library where the mail was kept. His eyes were wide with innocence as he said, "Oh, you mean Danes."

As a devout bachelor, Placzek was horribly afraid of the famous, unwritten Bohr Institute Rule: Any physicist working with Bohr was certain to be married after no more than two years. When his second year was nearly over, Placzek began to worry that his happy days as a single man were about to end. Since I was engaged to Ellen after only a year in Copenhagen, I decided to give him the year that was still owed me by tradition. In a solemn ceremony I transferred it to him so that he could be secure for one year longer. Much later when he did wed, the marriage was an unhappy one, so perhaps his instincts in this regard were sound.

In Russia (December 1936-January 1937)

I accepted an invitation from Kiev to go to Russia, to talk things over both with the Kiev people and with Kapitza. I did it because it gave me an opportunity to observe firsthand the changes that had taken place under the steadily increasing terror of Stalin, and I was eager to take Ellen on her first trip to the Soviet Union. As it happened, George Placzek was invited to the USSR at the same time, so we planned to travel together.

Placzek made a list of things that were hard to get in the USSR to bring to our Soviet friends. Among many other items he decided to include a large quantity of condoms. Since he had a very generous nature, he bought about a thousand of them. We decided to assemble all the goods in our apartment. On the appointed day, Placzek arrived with his arms loaded with packages. He began to unpack the boxes of chocolate, slide rules, tissues, and soaps he had bought. Then he suddenly stopped, hit his forehead with the palm of his hand, and said, "The Copenhagen taxi trade is going to die out!" We couldn't imagine what he meant. "I forgot the box with the condoms in the taxi," he finally told us, and he left to buy another huge batch.

We left for the USSR without any illusions, but what we found was even more dispiriting than what we had expected.

Summer 1938. Traveling in the US

In the summers of 1938 and 1939 I was invited to teach at Stanford

University. We drove across the continent, a great experience for us. There were no superhighways in those days, so we saw a great deal of the country intimately. Once in the badlands of North Dakota we got stuck in the mud in a wild area where there were no paved roads at all. Placzek came with us on one trip. In Copenhagen in 1934 we had made a bet about how long Hitler would last. Placzek; said not more than five years, and I said he would be around for much longer. At first we wanted to bet fifty dollars, but Placzek thought the value of the dollar was subject to change. Instead, he proposed, "Let's bet fifty Wiener schnitzel." At the time a Wiener schnitzel cost about a dollar. By the time we were on our way to Stanford, I had unfortunately won the bet. When we stopped at a restaurant that served Wiener schnitzel, Placzek threatened to pay off what he owed me by ordering fifty, but I managed to persuade him to pay me the money instead.

While we were out west, we visited Oppenheimer on his ranch in New Mexico. It was a wonderful place in the Pecos Valley near Los Alamos, which at that time was known to us only as the site of an obscure boarding school for boys. There was always a nice group of people at the ranch, and everyone pitched in with the cooking and the household chores. The ranch had few conveniences and no electricity, and we slept out on the terrace. Placzek, a cofirmed city dweller, said he couldn't sleep because chipmunks danced on his stomach all night.

In the summer of 1938 we had conversations with Oppenheimer that lasted all day and into the night. Oppie, as we called him, still believed to a great extent in communism. We tried to convince him of the reality of Soviet life by describing the lack of freedom and the persecutions under Stalin that we had seen there. I believe that these conversations had a lot to do with his eventual rejection of communism. He trusted us as stalwart liberals who were not ideologically anti-Soviet but who had had an opportunity to see life in the Soviet Union as it really was.

Cornell 1941

Although we were excluded from officially working on the nuclear bomb, Bethe, Placzek, and I discussed the problems involved in the project during my numerous visits to Cornell, and we tried to get some idea of what was involved in achieving a nuclear explosion. The slow and secretive way fission research was done at that time was very frustrating for us. Our fear that Hitler would develop the nuclear bomb before the Allies was a realistic one. After all, the fission process had

been discovered in Germany; and although many of the physicists had
left Germany because they were Jewish, the Germans still had Werner
Heisenberg and other excellent scientists.

Los Alamos

This is why the Theoretical Division was so important at Los
Alamos. We tried to get the best people available. Hans Bethe was
the division leader, and we had Enrico Fermi, Robert Christy from
California Institute of Technology, Edward Teller, Philip Morrison from
Berkeley, and later Rudolf Peierls from England. My old friend Placzek
came, and Robert Marshak joined us from Montreal, where he was
working on a British-Canadian project.

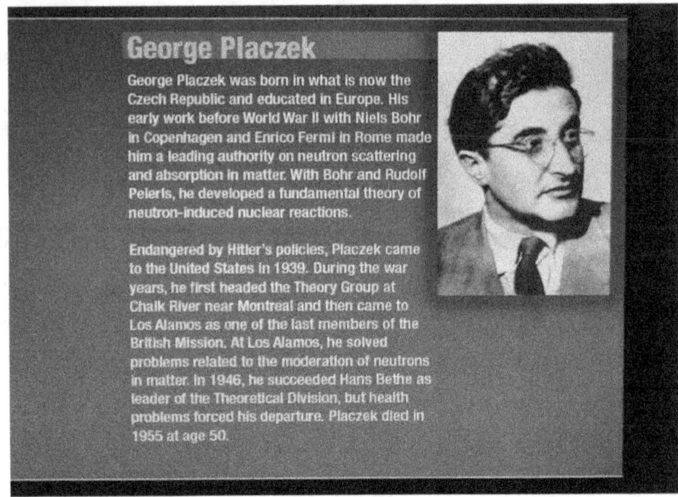

Information on G. Placzek in Los Alamos Bradbury Museum of Science.

Figure 5.4 Some documents describing work on nuclear weapon in Los Alamos are collected in this museum.

Excerpt from

Some Men and Moments in the History of Nuclear Physics: the Interplay of Colleagues and Motivations, by John Archibald Wheeler[13]

The barrier height of a compound nucleus against fission was not the only factor relevant for fission. Equally important in governing the probability of this process was the excitation, or "heat of condensation," delivered up by the uptake of a neutron to form the compound nucleus in the first place. On this point an important development occurred on a snowy morning when I was occupied with classes and not with Bohr. He, having breakfast at the Nassau Club with Rosenfeld and with an arrival of the night before, George Placzek, faced Placzek's continuing skepticism about the very existence of fission. Placzek asked, how can it possibly make sense that slow neutrons and fast neutrons cause uranium to split but not neutrons of intermediate energy? Bohr stopped but said not a word, left with Rosenfeld, crossed the campus to Fine Hall still without a word and there, when Placzek and I joined them, explained the great idea that had just come to him: that the slow neutron fission takes place in the rare isotope U^{235} and the fast neutron fission in the abundant isotope U^{238}. Thus an incoming neutron delivers up a high heat of condensation when it enters into a nucleus with 143 neutrons, because it can form a new neutron pair. This excitation puts the compound nucleus U^{236} over the barrier summit.Therefore, the U^{236} must split when it is formed by slow neutron capture. Moreover, the cross section for fission of the rare U^{235}, like the cross section for the "fission" of boron, $n^1 + B^{10} \rightarrow He^4 + Li^7$, must exceed by far the geometrical cross section of the nucleus for sufficiently slow neutrons. That circumstance makes it understandable why an isotope present to only one part in 139 imparts to natural uranium the observed substantial fission cross section. However, the cross section must fall off inversely as the velocity of the neutron for U^{235} as for B^{10}; hence the negligible fission cross section of natural uranium for neutrons of intermediate energy. In contrast, when a slow neutron enters U^{238} to form the compound nucleus U^{239}, no new neutron pair is formed. The heat of condensation delivered up is not enough to exceed the fission barrier. Only neutrons of a substantial kinetic energy striking U^{238} can produce

---

[13]From *Nuclear Physics in Retrospect*, Proc. of the 1977 Symp. on the History of Nuclear Physics in the 1930s, Ed. Roger Stuewer, (University of Minnesota Press, 1979), pages 276-278.

U^{239} with enough energy to surmount the barrier. Hence, the existence of fast neutron fission in natural uranium.

Placzek, wonderful person that he was, a man of the highest integrity, often a thoroughgoing skeptic about new ideas, said to me over and over in those early spring days of 1939 that he could not believe that the small amount of U^{238} could be the cause of the slow neutron effects in natural uranium. I therefore bet him a proton to an electron, $18.36 to a penny, that Bohr's diagnosis was correct. A year later Alfred Nier at Minnesota had separated enough U^{238} to make possible a test and sent it to John Dunning at Columbia to measure its fission cross section. On April 16, 1940, I received a Western Union money order telegram [14] for one cent with the one-word message "Congratulations!" – signed Placzek.

---

[14]G. Placzek, April 16, 1940, telegram, on deposit at the National Museum of Science and Technology, Washington, DC.

Excerpt from

British Scientists and the Manhattan Project, Ferenc Szasz[15]

Almost a third of the British Mission to Los Alamos – Bretscher, Frisch, Fuchs, Peierls, Placzek and Rotblat – were émigrés or refugees from the Continent. Thus, they were "British" as much by force of circumstances as by design. Peierls had become a British subject in 1940, but Bretscher and Frisch were naturalized, called up for military service, and then exempted within a twenty-four hour period just before departure to the States. Frisch actually received his British passport only after he had boarded the ship for the United States. In a unique arrangement, Joseph Rotblat remained a citizen of Poland while he worked on the Hill. These refugees, naturally, retained heavy continental accents, and a wag once remarked that the British Mission couldn't speak English.

Initially all the British scientists were to work under Chadwick, but he soon decided to leave Los Alamos for Washington. Chadwick's decision to leave was doubtless influenced in part by family concerns. Neither his wife, Lady Chadwick, nor their twin teenaged daughters adjusted easily to the Los Alamos environment.

Lady Chadwick had never been stateside before and experienced difficulty coping with the demanding living conditions at Los Alamos. Once she invited several wives over for a formal high tea and, during the course of the conversation, delivered an attack on the primitive nature of life in the United States. This annoyed several of the American wives.

---

[15]Ferenc Szasz, *British Scientists and the Manhattan Project: The Los Alamos Years*, (MacMillan, 1992), page 39.

Interviews

Excerpts from

Interview with Alexander Ilyich Akhiezer (1911-2000)[16]

Q. Landau had a [published] work with Placzek...

A. I forgot to tell you. Indeed, Placzek worked here, [in Kharkov]. He was here for quite a long time and had close contacts with Landau.

Q. Was he in the Theory Department?

A. No. It is hard to understand where exactly he was. He had close contacts with both, Landau and Obreimov.

Q. That is, you can't say to which department he formally belonged, can you?

A. Well... you cannot formulate your question in this way. Placzek was in a very close relationship with Landau. Moreover, when I was defending[17] my Candidate thesis[18] Placzek was on my defense committee.[19] All children around his dwelling knew him.

Q. It seems to me that Placzek was the Official Opponent of Tisza too...

A. Indeed, this was the case. All little children always teased him: "Placzek, Placzek, giv'me a *kalatchik*."[20] He was incredibly witty, one can rarely meet such a joker. He left before other [foreigners] were deported [or arrested].

Q. Got it. Did he work in Kharkov on light scattering?

A. Indeed, he worked on light scattering on molecules. It was a very important work. Then he moved on to neutron scattering and published a large paper on this in America. He acknowledged Pomeranchuk and me in this paper .

Q. OK, besides Landau, did Placzek collaborate with other theorists and experimentalists?

A. Yes, with Ivan Vassilievich Obreimov, who headed an experimental laboratory.

Q. Did he publish anything [resulting from this collaboration]?

A. I do not remember. You can check it if you look through all issues

---

[16]Conducted by Yuri Ranyuk. The full text is published in *Four Interviews*, Eds. Yuri Ranyuk and Yuri Freiman (NNTs KhFTI, Kharkov, 2010).

[17]In late December 1936 or early January 1937

[18]The Russian analog of Ph.D.

[19]In Russian Placzek was referred to as the *official opponent*.

[20]*Kalatchik* in Russia is a small round loaf in the form of a ring with a small hole, similar to American bagel.

of *Physikalische Zeitschrift der Sowjetunion.* If he published a paper it should be in *Zeitschrift der Sowjetunion.*[21]

Q. Was he married?..

A. He was as a bachelor here.

Q. And where did he stay? Was it the same building where other foreigners stayed? He was famous for his jokes, wasn't he? And the story with Weisskopf...[22]

A. As far as I remember he stayed in our hostel, but I am not sure. I hardly socialized with him outside the Institute. Placzek jokes... He was a remarkably talented man, and his jokes were numerous.[23] He was a friend of Pivovar,[24] talk to him, he will tell you.

$$*****$$

Excerpts from

Interview with Otto Frisch[25]

Frisch:
[...] The work which I did with Placzek was quite different. Bohr took a passionate interest in that, and he pressed us to publish at once and so on. He was obviously very keen on that. It was a glorious occasion when we actually wrote that paper at about two in the morning.

Placzek, I think, assumed that I would be too sleepy to finish the work because he didn't want to write it in such a hurry, but for once I actually managed to keep awake until four; and then Placzek issued a sort of declaration that under passionate protest he would allow me to take that paper to the mails. I took the paper to the mails at four in the morning. I wanted to make sure it wouldn't get delayed again the next day. And Placzek said, "Shall we phone Bohr?" And I said, "Well, Bohr is having dinner with the King." So Placzek said, "Let's look up the King." We looked up the Royal Palace in the phone book,

---

[21] I failed to find joint publications of Placzek and Obereimov. -MS

[22] See page 291.

[23] According to Vladislav Syshchenko, Akhiezer's student, Akhiezer liked to tell his students of Plazcek's joke about his fear of radiation. Each time Placzek was passing near a certain laboratory on the UPTI campus, he would exclaim, "Oh, I am afraid of radiation," and would pretend that he was covering his genitals with his palms.

[24] Lev Iosifovich Pivovar (1910-2000) was the Director of the UPTI Central Experimental Workshop.

[25] Courtesy of AIP,
https://www.aip.org/history-programs/niels-bohr-library/oral-histories/audio/4616

but there were so many phone numbers we didn't know which one to phone.

*****

Excerpts from

Interview with Victor "Viki" Weisskopf (1908-2002)[26]

Victor Weisskopf. We had a little party and we thought... There was George Placzek, they wanted him to take a job. He had a sense of humor, and they thought there were all good friends. Placzek said he had five conditions. The first condition is a good salary, and third of the salary in dollars.

F. Yes, it was a condition for his...

W. For his going to...

F. To Kharkov.

W. The second condition was the laboratory with two assistants, and the third the support for his research. The fourth condition was that he can be three months a year away. And the fifth condition: "khazain" must go, the "khazain" must go. The "khazain," that is Stalin.

F. Oh! It was very dangerous joke!

W. Yes, but he thought... And Barbara Ruhemann told this to the party. He was invited by Alex Weissberg and then Weissberg was arrested.[27] Later he played a role in the Polish underground, a very

---

[26]Conducted by Yuri Freiman in 1990. The full text is published in *Four Interviews*, Eds. Yuri Ranyuk and Yuri Freiman (NNTs KhFTI, Kharkov, 2010).

[27]The joke of Placzek as narrated by his close friend, Edoardo Amaldi, from *The Adventurous Life of Friedrich Georg Houtermans, Physicist* (Springer, Berlin, 2013), Eds. Saverio Braccini, Antonio Ereditato, and Paola Scampoli, (pp. 40-41):

"Once Martin Ruhemann, a German physicist working in low temperature, gave a reception for all physicists present in Kharkov, Russian as well as foreigners.

Not long before, Placzek had received the offer of a permanent chair, and a few of the people present asked him what was his decision. Placzek, in his typical joking mood, answered that he was ready to accept provided five conditions were fulfilled. "But what are the conditions?" his friends asked.

The first condition was that his salary should be larger than a certain satisfactory value, the second one that about one third of the salary had to be paid in pounds because he wanted to spend every year two or three months abroad in order not to lose contact with the international scientific community. The third condition was that at least two young people who could work with him had to be paid in some form. The fourth condition was that the economic treatment of these young people had also to be decent. Finally, the fifth condition was that the "Khozyain must go."

"Khozyain" in Russian means Boss (or Master) and at that time was an euphemism for Stalin. Everybody laughed and commented humorously.

The wife of Ruhemann, Barbara, who was a very rigid party member, reported this little story to

important role.

F. Yes. You met him...

W. Oh, I knew him [well]. He was from Vienna. He also organized my invitation to Kharkov. He was the managing director of the Physical Technical Institute in Kharkov.

F. What do you mean when you said 'the managing director'? Was this position...

W. Well, he was not scientific director. He was a physicist, but he essentially did administrative work.

F. And he offered you position in Physical Technical Institute, right?

No, that was Placzek. He invited me. I had time, I had eight months between two positions and I wanted to go to the Soviet Union because half of my friends said it was wonderful and half of my friends said it was terrible.

F. It was terrible.

W. I wanted to find out what is was. And I found out it was terrible. Even at that time. I wanted to stay for a long period, for eight months.

F. Was it an offer from Weissberg?

W. I asked him because I wanted to go to see the Soviet Union. He arranged my visit. We had a room. I came with a girlfriend, you know, she was not my wife. I was just given a room and board, and I gave lectures. There was not much to eat anyway.

F. And your girlfriend, was she Austrian?

No, she was from Berlin, her name was Ruth Bonaril, and she actually fell in love with a Russian, by the name of... I think Ginsburg... no... Not a physicist. A man, a political man. And a Jew. And then he was killed by Stalin. First he was ambassador or something like that in Urumchi in Western China. There they had a child, they had a primitive life and then in the чистка (purges) he was killed, and she had a very unhappy life with the child. She was a photographer. Now she lives in East Berlin. But when I left she stayed in Russia.

No, that's not right... She came back to Berlin and then returned

---

the Communist Party in Kharkov with the result that shortly after, the local newspaper started to attack the foreign physicists with gradually heavier and more explicit accusations of being German spies.

   Weisskopf, Placzek and most of the others understood immediately the general trend taken by the situation and rapidly left the USSR. Weisskopf went to Zurich where he stayed three years, married Ellen Tvede in 1934, and after a short return to Copenhagen went to the USA in 1939.

to Russia.

F. OK. When exactly did you arrive in the Soviet Union for the first time?

W. That was from... I don't know exactly... From April to October or something like that.

F. And what position did you hold?

Oh, no position! I mean I gave lectures and worked in Physical Technical Institute with Landau.

I must tell you one interesting episode. One day we came to our room and saw that everything had been stolen, everything, all the clothing had disappeared. And also of my girlfriend's... Weissberg wasn't there. I informed the police, you see, and nothing happened. Weissberg came back about week later. When he came he asked me: "Well, how is life in the land of workers and peasants?" He was very communist.

F. Yes, I know.

And I said: "Terrible, as they stole all my belongings." Then he said: "I'd make a bet with you." I don't know how much, 50 dollars or so. "If you get everything back in two days." And I said: "That's ridiculous, it's already more than a week and the police didn't even try to..." He said: "You'll get it back." I said: "Fine." And – "Give me a list," he said. We gave him the list. And indeed in two days we got everything, clean, nice... So I paid him my bet. And I asked him: "But how come?" He said: "That's very simple. They... You know the KGB has informers in different bands of thieves. I went there and I said it's very important to have a good reputation abroad and that you are an important scientist. They notified all informers, and they brought everything back." There was everything back to my list, and even an overcoat that I forgot to put on the list, you know. That is an interesting story.

F. Yes, of course. You said the conditions of you life were not...

W. Well-well, there was very little to eat and we had only one meal a day and...

I remember the following thing. We asked Weissberg, a group of us, I myself, a friend of mine and then another, Motz, and Éva Striker.[28] You know her, right? We wanted to see how *kolkhozes* worked.

She was for a certain time the wife of Alex. She was there too. We four wanted to visit a *kolkhoz*. And since we didn't speak Russian very

---

[28]See Fig. 5.5 referring to 1934. Courtesy of Jean Richards.

well we went to a German *kolkhoz* because there were German people at that time, German regions. So we went to a German *kolkhoz*. In Vysokopolie. It was a village south of Kharkov, pretty far south. Do you know the name? No? It's a small place.

We took a train there. Finally we arrived there, and the chief of that *kolkhoz* was an Austrian communist and so we were celebrated as the proletarian brothers from Vienna, you know, and treated extremely well. They... There they had a lot to eat because... You see, the main reason why the *kolkhozes* did not work was because the instructors from Moscow didn't understand anything and the peasants didn't know what to do. But in these German parts the education was higher and they actually were quite successful. But only were a few. And... So... There is also an interesting episode. There was Sunday and our room was just in front of the church. By the way that's all in my book in my autobiography, you could read it there. And... There was a church, that was I think... German and Lutheran...

We looked at it and then two of my colleagues – I was a sort of the leader, I was the oldest, – two of my colleagues, Éva Striker and Motz, they wanted to go to the church. And I said: "Better not," – but they went anyway. And then later on I got a telephone call from the leader, and he said that was a very serious thing and I should come to him for a conversation. And I went to him. And he said we did a very bad thing – two of us went to the church, our anti-religious propaganda is... I said... And in the afternoon of that Sunday there was a big celebration for us. And I should have given a speech. He said I should make a remark that this was not because we believed in religion but only to find out how many people go to church, you know, how many people went to church. So I did give that speech and said we are not satisfied with the anti-religious drive, you know... And then everything was OK in this aspect. There were many funny stories. For example, when we were there, you know, they showed us all the agriculture what they were doing. And then we went together and then we had to make a critical summary. We had to say what did they do wrong. We were scared. That was very difficult for us: what did we understand? And Éva Striker, Alex's wife, she was actually very good, she said we were at the kitchen and we saw the people peeling potatoes cut off big pieces of potatoes, and that was a waste. Eva said was they ought to boil the potatoes first, and then peel them. That way they would save more potato. So we had at least something to criticize. And then there was a big celebration, dances, and we were brought to the station with great celebration, so that was our visit to the *kolkhoz*. But of course this *kolkhoz* was by no means characteristic, in other Russian *kolkhozes* nothing worked.

I remember one more thing. Every week we had Marxist lessons where we had to go.

F. In the Institute?

In the Institute. And some idiots had talks, you know, just silly, you know, *kagebists*...

F. I can imagine.

You can imagine. Landau and I were sitting in the last row. Landau made always dirty remarks, and quite loudly. I was afraid there.

F. Did many people attend?

W. Oh, they had to. All physicists and engineers. One could not stay away. Except if one had an official excuse...

F. I know it from my own experience.

W. Sure. It was absolutely idiotic. Even if you believed in Marxism, it was idiotic. Absolutely, absolutely, senseless. Landau made ironic remarks all the time, and rather loudly, and I was very much afraid. He was arrested later, but not at that time.

Figure 5.5   Eva Striker in the USSR, 1934.

Three years after Stalin died we resumed visiting Russia and Russians came here. I was very active in inviting Russians to Washington. I visited Russia in 1956, it was 20 years that I didn't see Landau. It was great, we embraced each other, it was wonderful. And then I went to Landau's home, I went to the washroom to the toilet and there instead of toilet paper was Stalin's biography. I took one page as a souvenir...

*****

Excerpts from
Interview with László Tisza[29]

Q: The Kharkov conference of 1934 could be a good start for our conversation. Was it your first visit to Russia?

A: Yes.

Q: May 1934. Who invited you?

A: Landau invited me at the instigation of Teller and Placzek. *May be it was Placzek. He was closer to Landau.*

Q: Did you see Landau before?

A: No.

Q: So, it was your first look at the communism experiment to which you felt some sympathy. As the majority of your generation, I believe? I am not sure that Teller was sympathetic then?

A: No, he was not. He was a good friend of mine. He was very anti-communist.

Q: I thought he became anti-communist by the late 30s when he realized that Landau, an ardent communist, had been arrested?

A: I talked to him in 1936 during my second vacation from Kharkov. I arrived to work in Kharkov in January 1935. And the first vacation I spent in Hungary. The next year I decided to spend some vacation in Austria, in Austrian mountains. Teller and his wife were on their way back to stay in the United States and we arranged our meeting. At that time I told him how much I changed my view.

Q: So, you changed your mind by the summer of 1936?

A: Yes.

Q: I think, that Landau didn't change?

A: Not yet. When I was back in Kharkov in August 1936 I learned the news of the execution of Zinoviev and Kamenev. But Teller was biased against communism back in the early 30s, when I told him about my sympathy to communism.

---

[29]Conducted by G. Gorelik, February 28, 1998; authorized May 28, 1999.

Obituaries

Hove, L. van

GEORGE PLACZEK

1905 · 1955

INSTITUTE FOR ADVANCED STUDY

PRINCETON, NEW JERSEY

Van Hove's Oration title page, published by the Institute of Advanced Study, Princeton.

Figure 5.6   George Placzek's photograph (circa 1950) from the published text of the funeral oration for George Placzek, pronounced by Léon van Hove on October 14th, 1955, in Zürich. (Courtesy of the IAS Library.)

GEORGE PLACZEK [30]
1905 · 1955

Léon Van Hove

These words are the funeral oration for George Placzek, pronounced by his friend and colleague, Professor Léon Van Hove, on October 14th, 1955, in Zurich. They are printed here because they speak with precision and eloquence for Placzek's friends at this Institute and throughout the World.

The painful moment has come to us, brought together here by the sentiment of shared affliction, to say the last farewell to the man, to whom we were attached, probably, by various ties, but also by the very admiration for his personality, his knowledge and his work. The loss of George Placzek, carried off from his relatives on the very threshold of his fiftieth year, can be measured only if we think about all the intentions and all the knowledge that he had in himself and about all the use that he could yet make from them if he was given a longer life. His published works were sufficient to give him a choice place in physics of the century, and every competent physicist nowadays is familiar with his name. But only several privileged individuals, who were able to notice, during long conversations, the scale of his knowledge and the depth of his judgments, can realize the role that he could have been playing for many years if illness and death had not come to interrupt him.

Despite being too short, the life of George Placzek was particularly active and very fulfilled. Few people realized as clearly as he did, in our century of nationalism often narrow, the old ideal of a scholar traveling all over the world, staying successively in the largest foyers of scientific creations with the purpose to study, produce and convey knowledge. George Placzek did his university studies very young, first briefly in Prague, and then in Vienna. He devoted himself immediately to research in the field of physics and, very early, his work turns to the questions that will remain dear to him for more than 25 years: the theory of scattering of radiation and of particles in matter, a field of work with various and often difficult aspects, in which he will become one of the very rare experts. After several years spent in Utrecht and a short teaching stint at the University of Jerusalem, he works – we

---

[30] Also published in Nuclear Physics, 1, 623 (1956). Translated from French by Valentina Maslova, Sean Kalafut, and Maxim Konyushikhin.

are at the beginning of the 1930s – in Copenhagen, at the Bohr Institute. He also stays several times in Rome and there participates in the work of Fermi and his group. Thus he observes on site, in Italy and in Denmark, fundamental works on nuclear reactions caused by neutrons. Death unfortunately came to interrupt him before his works could bear their most beautiful fruits.

The published work of George Placzek contains a significant number of articles, a subject that quickly acquires a primary importance for the subsequent development of nuclear physics. Still between 1930 and 1940, George Placzek travels several times to the Soviet Union and there accomplishes with Landau important research on light scattering, mostly not published and of which very little is known. He also goes to the United States and there stays at Cornell University. In the meantime, he works at Debye's place in Germany, and at the Henri Poincare Institute in Paris. From the beginning of the Second World War, he participates in scientific research demanded by military investments, first in England, then in Montreal, and finally in 1945 in the Los Alamos Laboratory. He plays a fundamental role in the theoretical work required by the construction of the first nuclear reactors and remained ever since, for these questions, a consultant of United States Atomic Energy Commission. After a period of work at the General Electric Company, George Placzck finds finally in 1947, at the Institute for Advanced Study of Princeton, a disinterested harbor of science and research, which allows him to plunge again, after the turbulent years of war, into the work of his choice. He develops there the theory of neutron scattering in crystals and also delivers the necessary foundations for the interpretation of the growing category of experimental works. Death unfortunately came to interrupt him before his works could bear their most beautiful fruits.

The published work of George Placzek contains a significant number of articles and two large scale works: a monograph on the scattering of radiation by atomic and molecular systems, which appeared in the "Handbuch der Radiologie," and a treatise on the multiple scattering of neutrons in matter, based on a series of lessons given by the author at the Los Alamos Laboratory.

Arriving at the end of this quick outline of the life and works of the deceased, I cannot help to say to myself how this enumeration gives such an incomplete idea of who was the man that we mourn for. Those who have approached him and had thus a chance to learn little-by-little the true aspects of his personality, were all touched by a delicate, sensitive man, deeply educated, who was shrouded in harshness and skepticism. Few people had an appearance as deceptive as George Placzek, and that's why so often people, on the basis of impressions, collected through a meeting too brief, made an opinion of him whose superficiality and falsity was evident to his friends. George Placzek, not much inclined to trivial conversations, did not open up easily, and developed new relations slowly. In addition, since the war, the slowly growing certainty that his closest relatives died tragically, did not fail to darken his character and close him in on himself.

And yet, what treasures and what shades did his personality unveil to those who had learned to know him better? The extent of his intellectual faculties, already so striking for those who knew his scientific works, disclosed in other days his amazing knowledge of languages, history, and literature of several countries. It is clear that George Placzek read enormously, and had a rare faculty to remember facts and at the same time to make a very personal judgment of them. To give an example, he was interested in the history of the United States, which became his adopted homeland since the war. He amassed such a complete knowledge that many Americans born in the country envied him, and one thought that he could often reveal the true meaning of the present political events of this country, which is so difficult for Europeans to understand. And to characterize the diversity of interests of the deceased, I will content myself with citing his affinity for the literary and scientific works of the Middle Ages and Renaissance, to which his perfect knowledge of Latin and of Italian gave him immediate access. Last May, in Paris, though already in bad health, he was telling me with visible pleasure about his project to give a lecture in Rome this autumn on one treatise of geometry from the 14th century, forgotten in the Vatican library and not studied until this day.

So who, among friends of George Placzek, could ever forget the quality of emotions that he was giving to them? His joy, always so spontaneous, to see again his friends, and the cheerful attention with which he worried about their fate and strived to help them with his advice, were probably surpassed only by his profound and unwavering respect for the convictions of others, a fundamental and nearly instinctive tolerance, which often pushed him to justify the point of view of others with more eloquence than his own. These qualities of a great heart, joined with a lot of perceptiveness, with sharp critical judgment and with such diverse life experiences, made him an ideal man to see through complex situations and to wisely make difficult and significant decisions. Also numerous were friends of George Placzek who were coming to consult with him when a tricky problem, sometimes a decisive turning point, was facing them. They would find him always ready to help them, and – I am speaking from personal experience – his interest and his benevolence were so spontaneous that people often used to exploit his time and his patience a little more than was necessary. I know more than one physicist who will feel later, in one or another difficult circumstance, how hard it is to make a decision when George Placzek is not here to light the way.

This critical and very certain sense, these qualities of perceptiveness and objectivity, of which so many friends of George Placzek profited from, were also – this comes without saying – the remarkable characteristics of his scientific work. One more reason why they should be emphasized is that too many researchers of the post-war generation seem to fail to recognize their fundamental importance. Too few young people had an opportunity to closely observe the work methods, reasoning, and critical review of the results used by scientists of such a high caliber as the deceased, and here also we see how much eminent help he could have produced if illness did not destroy him. This mercilessly critical mind, this desire for perfection that lead him in the review of his own results, was also applied to the works of others. We could draw up a long list of problems to which he rebuilt the solutions proposed in the literature on a more satisfactory basis. To a great extent, the product of this work is unfortunately lost and, as I am seriously afraid, the same destiny is reserved for more than one result obtained by George Placzek personally. As much as he was putting energy and patience to perfect and to chisel his methods and his results, putting these on paper and the final redaction were heavy burdens on him. His interest in the problem fell as soon as he had found a solution in several scattered notes, and often he could not help but to extend his research further before he recorded the obtained results. Here also we can measure the loss which implies his premature death.

Gathered on this sad occasion, we cannot impede ourselves to group thus in our memory so many remembrances, so many aspects of the personality of George Placzek, and each amongst us most painfully feels the weight of his loss. We remember his sufferings of the last months, when the qualities of perceptiveness and of realism, which were his force, turned against him and made him see in the full face the tragic character of his state. We assess the pain that hurts you, Els Placzek, even to a larger extent, his daily companion, you looked after him, supported him, encouraged him, with such patience, without being given, because of another tragic bereavement, the opportunity to be with him at the supreme moment. Now, as death has come to relieve his sufferings, try to think that all the reasons which render his loss so cruel, also make his life worth living. All the qualities, all the traits of his character which made him so dear to you, his wife, and also to his relatives and to all his friends, who are unfortunately far away, make perennial his memory and it will be a rich source of inspiration and courage for those who live on. And the certainty to the above said about George Placzek, a good man and a scientist of high class, a person at the same time human and deeply realistic, is the most beautiful word of farewell which we can address to him.

George Placzek

Edoardo Amaldi [31]

On October 9, 1955, George Placzek died in Zurich some months after the return of a serious illness, which had troubled him since his arrival in Europe in late spring of the same year. He had come to Europe to stay for a lengthy period, promising himself to spend the academic year 1955-56 at the Institute of Physics of Rome University, thanks to a grant from the Guggenheim Foundation.

Although his friends had known about the state of his health for a number of years, they had still hoped for news of his recovery at a Zurich clinic, as had happened on previous occasions. The news of his passing was so unexpected causing great sorrow for those who knew him, in particular the physicists that best had the opportunity to appreciate his humanity and scientific talents, but especially the Italian physicists with whom he had been closely associated for many years.

George Placzek was born on the 26th September 1905 in Brünn, Czechoslovakia,[32] and studied in Prague, then Vienna where he obtained his PhD in physics in 1928.

From that moment onwards, he commenced a period of travels around important physics research centers in Europe. From 1928-31 he was in Utrecht where he studied with Kramers and with whom he developed a great friendship. During 1931 he was in Leipzig, at the Institute of Debye and Heisenberg. In 1931-32 he was in Rome where he formed scientific and personal contacts within the group of young people working under the supervision of Fermi. This would bring him back to the University of Rome for lengthy periods, every two or three years, for the rest of his life.

During 1932-33 and 1935-38 Placzek worked at Niels Bohr's Institute of Theoretical Physics in Copenhagen; in 1934-35 he was nominated Professor of Theoretical Physics at the University of Jerusalem, and in 1933 and 1936-37 was a visiting professor at the University of Kharkov [33] where he worked with Landau.

After a period at the College de France of Paris in 1938, Placzek moved to the United States, where he was professor at Cornell

---

[31] This obituary was published in *La Ricerca Scientifica*, N°7, pp. 2037-2042, Luglio 1956. Translated from Italian by Stefano Bolognesi.

[32] Currently Brno, Czech Republic.

[33] More exactly, Placzek's host institution in Kharkov, Ukraine, was the Ukrainian Physics and Technology Institute, UPTI.

University, Ithaca, NY, from 1939 to 1942.

When the war started, Placzek joined, with an important managerial role, the Theoretic Division of the research group in Applied Nuclear Physics organized by the British Authority in Montreal, Canada. He stayed there until 1945 when he moved to Los Alamos Laboratory in the US. After the end of the war, he spent a period of almost two years at the General Electric Laboratories in Schenectady, leaving in 1948, when he was called to the Institute of Advanced Studies in Princeton where he remained until his death.

His scientific contributions cover many problems in physics, from questions in molecular physics to problems in nuclear physics, both applied and fundamental.

A number of works, undertaken in the period from 1929 to 1934, dealt with the quantum theory of the Raman effect. Placzek was the first to study in a systematic way the relations between diffusion of light in molecules and its symmetry properties, and was able to formulate an elegant general theory that constitutes the basis for all future works on the subject. This theory has been summarized in a fundamental paper entitled: "Rayleigh Streuung und Raman Effekt," published in the second edition of *Handbuch der Radiologie* [15]. The uses/applications of these general methods allowed him to solve some important problems, such as the diffusion of light off vapors at the critical point, liquids and crystals. Additionally, he was able to gain, since his young years, an exceptional ability in the methods of investigation of the diffusion of radiation off small bodies from many different systems, an ability that would made him one of the greatest experts of our age in this field.

His frequent contact with the groups in Rome and Copenhagen directed his attention around 1934-35 toward nuclear physics. During his stay in Copenhagen in 1936, a few days after Niels Bohr published his nuclear model explaining the resonance recently found in the absorption of slow neutrons, Placzek published a letter to the Editor of *Nature* together with Frisch [16]. In this letter argumentation was presented that led to attribution of the well-known factor – inverse proportionality to velocity – the capture cross section of slow neutrons by boron and lithium. This relation, known as the $1/v$ law, at that time became a very important comparative tool for the solution of many problems in the field of "spectroscopy of slow neutrons" that was just starting. In a few years the above relation was verified experimentally with great accuracy thanks to the use of precise spectrometers.

The following year, together with Bethe, he published a fundamental work on the resonance effects in nuclear processes [17] in which Bohr's ideas and the formalism used by Breit and Wigner to derive their "one level formula" were used and elaborated in a general theory which represents, even now, a great step toward theoretical understanding of nuclear processes.

In 1939, together with Bohr and Peierls, he published in Nature [18] a short letter to the Editor on the so-called "optical theorem" which specifies the absorption and diffusion of fast neutrons off nuclei, a problem on which the same authors wrote a more detailed paper that was shared privately between a small personal circle, but never published due to the war-time circumstances which separated the authors.

A vast and important group of research dealt with the slowing of neutrons that he studied both before the war, concerning the neutrons in the cosmic radiation [19], then during and after the war in relation to his activity toward the computation of the moderator [20,21,22,23,24,26] and the multiplication of neutrons [25] in nuclear reactors. These works, partially done in collaboration with young students, are often characterized by clear and detailed solutions to difficult and delicate questions which others had not previously considered.

At least a part of his results mentioned above are collected in a volume published in collaboration with Case and de Hoffmann under the title *Introduction to the Theory of Nuclear Diffusion* (Vol. 1), see [31], which was based on the course given by Placzek in the summer of 1949 in Santa Monica, California.

After the war he further refined the analysis of the processes of elastic and inelastic diffusion from liquids in crystals, both in connection with the problem of establishing the existence and the magnitude of the neutron-electron interaction, from the results and experiences of nuclear diffusion [29], and both to develop appropriate methods of analyzing the structure of the liquids and crystals themselves [27,32,33,34].

In this last important works, he collaborated with Van Hove, who can be considered as his student in this field, and his last great friend. Even from this brief examination of the works published by Placzek it is clearly seen how important his contribution was for the development of various fundamental chapters in contemporary physics. Nonetheless the works published are not sufficient to give a complete idea of Placzek the scientist.

While he was always ready to study in depth, trying to uncover complex nature, Placzek was lazy in writing down and publishing the results he had obtained. Writing a paper was an enormous effort for him, to the point that often in his life he ended up not publishing his clearly established result. For instance, he published [14] only a part of his work undertaken with Landau during the period spent in Kharkov.

This aspect of his character was due in part to his desire to further develop analyses of the phenomena under consideration, a desire that often showed up as a dissatisfaction with what he had already obtained, or with the formulation he or his collaborators were able to give. Gifted, with a lively critical spirit, he was ready to understand the problems he addressed and outline all their nuances. These qualities, combined with his unselfish generosity, made him a precious, or, better to say, incomparable adviser.

These characteristics of Placzek the scientist were simply a reflection of his qualities as a man, his general culture outside of the field of physics which was incredibly vast, and profound, and solidly anchored in a classical basis. He not only spoke and wrote fluently in many languages – from Russian to Spanish, from Danish to Italian, from Dutch to French – but he also knew the history of the respective countries of which he read in their original languages, as well as both, classic and modern literature. For instance, once he moved to United States he learned in an uncommon way the history of his country of adoption, acquiring knowledge of details known only to academics.

His curiosity to always learn and understand new facts of the human life, big and small, made him a typical European man. In early youth Placzek had absorbed so many apparently diverse and contradictory skills that it seemed he should have belonged to all various populations residing in the Austro-Hungarian Empire.

The tragic destruction of his family by the Nazis during the war and the political events in his country, made him somewhat distant from his homeland, of which he spoke only with close friends and only if asked, but did not distance him from Europe. In the later years he hoped for an economical and cultural revival in Europe. Additionally, when he spoke about Europe he often referred to small countries, such as Denmark and Holland, where he spent time in his youth, and to which he was connected through his Dutch wife Els. Yet mostly he referred to Italy, which he had great affection for, not only appreciating the virtues but also recognizing the imperfections. He knew our literature and our history like few Italians do, he sang popular songs from various regions of the country and used to tell to his students histories with fine and acute spirit. After the war he greatly assisted in all possible ways to aid the recovery of physics in Italy, either supporting the young students who wanted to travel to the United States through grants, or trying to teach them himself what he knew so deeply. His penultimate trip to Europe was in the spring and summer of 1953, and it was devoted to short courses on neutron moderation and diffusion at the universities of Rome and Milan. He subsequently spent a brief vacation in Riccione and Venice, a time that both he and his wife remembered with joy.

He wanted to stay in Italy longer, so he applied for a Guggenheim grant that allowed Els and him to live and work in our county for many months during the academic year 1955-56. They were attracted to Rome, not only because of the old friends there, but also because of the historic and artistic value of the city that both of them knew and appreciated, but from different point of views: George focussed on art galleries and picturesque corners of the city, life in modern and old neighborhood of Rome, on rare books he was able to find in the libraries of Rome and Vatican.

He was for many years a member of the Italian Society of Physics, whose existence and development he was interested in; he was greatly pleased when a friend told him he considered Placzek to be an Italian physicist.

It is also for this reason and from this viewpoint that today we want to remember his name and his character with profound respect.

References [34]

This bibliography does not pretend to be complete, but certainly contains the most important works of George Placzek.

[1] "Ponderomolorische Wirkungen des Lichtes auf Getändene Submikroskopische Körper im Elektrischen Felde", *Zs. f. Phys.*, 49, 601, 1928.

[2] "Zur Theorie des Ramaneffektes," *Zs. f. Phys.*, 58, 585, 1929.

[3] "Zur Dichten und Gestaltbestimmung Submikroskopischer Probekörper," *Zs. f. Phys*, 55, 81, 1929.

---

[34]See also page 238.

[4] "Ueber die Lichtzerstreuung beim kritischen Punkt," *Phys. Zeits*, 31, 1052, 1930.

[5] "Ueber den Ramaneffekt beim kritischen Punkt," Proc. Amsterdam, 33, 832, 1930.

[6] "Polarisations messungen am Ramaneffekt von Flüssigkeiten," (with W. R. van Wijk), *Zs. f. Phys.*, 67, 582, 1931.

[7] "Intensität und Polarisation der Ramanschen Streustrahlung mehratomiger Moleküle," *Zs. f. Phys.*, 70, 84, 1931.

[8] Contribution to Leipziger Vortrage, 1931, page 71.

[9] "Ueber das kontinuierliche Ramanspektrum und sein verhalten beim Kritischen punkt," (with W. R. van Wijk), *Zs. f. Phys.*, 70, 287-292, 1931.

[10] "Evidence for the spin of the photon from light scattering," *Nature*, 128, 410, 1931.

[11] "Ramaneffekt des gasförmigen ammoniaks," (with E. Amaldi), *Naturwissenschaften*, 20, 521, 1932.

[12] "Ueber das Ramaspektrum des Gasförmigen Ammoniaks," (with E. Amaldi) *Zs. f. Phys.*, 81, 259, 1933.

[13] "Die Rotationsstruktur der Ramanbanden mehratomiger Moleküle," (with E. Teller), *Zs. f. Phys.*, 81, 209, 1933.

[14] "Struktur der unverschobenen streulinie," (with L. Landau), *Physikalische Zeitschrift der Sowjetunion*, 5, 172-173, 1934.

[15] "Rayleigh Streeung und Raman Effekt," article in *Hdb. der Radiologie*, 1934, Vol VI, 2, p. 209.

[16] "Capture of Slow Neutrons," (with O. Frisch), *Nature*, 137, 357, 1936.

[17] "Resonance effects in nuclear processes," (with H. Bethe) *Phys. Rev.*, 51, 450, 1937.

[18] "Nuclear Reactions in the Continuous Energy Region," (with N. Bohr and R. Peierls), *Nature*, 144, 200, 1939.

[19] "Interpretation of neutron measurements in cosmic radiation," (with H. Bethe and S. A. Korff), *Phys. Rev.* 57, 573, 1940.

[20] "On the Theory of Slowing down of Neutrons in Heavy Substances," *Phys. Rev.*, 69, 423, 1946.

[21] "The spatial distribution of neutrons slowed down by elastic collisions," *U. S. Atomic Energy Commission*, MCDD-2, 66 pp., 1946.

[22] "The concept of albedo in elementary diffusion theory," (MIT nuclear Science and Engineering Seminar 39), *U. S. Atomic Energy Commission*, NP-160, 16 pages, 1947.

[23] "Milne's Problem in Transport Theory," (with W. Seidel), *Phys. Rev.*, 72, 550, 1947.

[24] "Angular Distribution of Neutrons Emerging from a Plane Surface," *Phys. Rev.*, 72, 556, 1947.

[25] "A theorem on neutron multiplication," (with G. Volkoff), *Canad. J. Res.*, A, 25, 276, 1947.

[26] "Theory of Slow Neutron Scattering," *Phys. Rev.*, 75, 1295, 1949.

[27] "Effect of Short Wavelength Interference on Neutron Scattering by Dense Systems of Heavy Nuclei," (with B. Nijboer and L. Van Hove), *Phys. Rev.*, 82, 392, 1951.

[28] "Correlation of Position for the Ideal Quantum Gas," Proceedings of the Second Berkeley Symposium on Mathematical Statistics and Probability, 1950: 581-588 (1951).

[29] "The scattering of neutrons by Systems of Heavy Nuclei," *Phys. Rev.*, 86, 377-388 (1952).

[30] "Scattering of X-Rays by Atoms," *Phys. Rev.*, (2) 86, 588, 1952.

[31] *Introduction to the Theory of Neutron Diffusion*, vol I, (with K. M. Case and F. de Hoffmann), Los Alamos, Los Alamos Scientific Laboratory, 1953, 174 pp.

[32] "Crystal Dynamics and Inelastic Scattering of Neutron," (with L. Van Hove), *Phys. Rev.*, (2), 93: 1207-1214, 1954.

[33] "Incoherent Neutron Scattering by Polycrystals," *Phys. Rev.*, (2) 93, 895-896, 1954.

[34] "Interference Effects in the Total Neutron Scattering Cross-Section of Crystals," (with L. Van Hove), *Il Nuovo Cimento*, (1), 233-256, 1955.

GEORG PLACZEK, Theoretical Physicist

Victor F. Weisskopf[35]

Department of Physics, Massachusetts Institute of Technology, Cambridge

Last October George Placzek died in Zurich at the age of 50 after a long and painful illness. To a devoted circle of friends his death came as a terrible shock. To the world of physics it means the loss of a theoretical physicist with a universal and lucid understanding of physics, for whom the well-founded structure and significance of a theory were more important than the immediate adjustment to some recent observations.

Placzek was born in Czechoslovakia and began his scientific work in Vienna, later spending time at many of the scientific centers in Europe. He worked in Holland with Kramers, in Rome with Fermi, in Copenhagen with Bohr, in Russia with Landau. In 1938 he came to the United States and worked at Cornell University until the outbreak of World War II. He participated in the war effort, first in Montreal with the Canadian Uranium Project and later in Los Alamos; after the war he worked at the General Electric Research laboratory for about 2 years and then spent the rest of his life at the Institute of Advanced Study at Princeton University.

Placzek's contributions to physics range over many fields. Most widely known is his work on the Raman effect during the period 1929-34. He put the theory of the Raman effect of molecules on a new basis by incorporating the classical description (as proposed by Cabannes and Rocard) into the quantum formalism. An entire science is based on the use of the Raman effect for the determination of molecular structure; the pioneer work and the fundamental ideas of these methods are mainly due to Placzek. He was the first to investigate systematically the relationships between the scattered light of a molecule and its symmetry properties. His development of the theory of scattering by molecules is a masterpiece in its generality and intrinsic beauty. It can be found condensed in a review article by him, "Rayleigh Streuung und Raman Effekt," in the Handbuch der Radiologie, ed. 2, vol. 6 (1934). His studies of the scattering of light enabled him to solve a number of problems in this field with better and more general methods than had been used before. Examples are a study on the scattering off media at

[35]Published in *Science*, March 9, 1956: Vol. 123, Issue 3193, pp. 409;
DOI: 10.1126/science.123.3193.409

the critical point and on the scattering of crystals and liquids. In this period he acquired a mastery of the problems of scattering that made him the foremost expert in this field.

In the early 1930s, Placzek spent some time in Rome with the group around Fermi and in Copenhagen at Niels Bohr's Institute. He was attracted by the newly developed neutron researches and worked experimentally and theoretically at the exploration of the fascinating problems of neutron-induced nuclear reactions. With O. R. Frisch, he published some work on the capture of slow neutrons, and, after coming to the United States for permanent residence in 1937, he joined with H. A. Bethe in the fundamental paper on neutron resonances, which gave a strong impetus to the development of our knowledge of slow neutron reactions. From then on his interest remained focused upon neutron physics. His great experience in the theory of scattering that he acquired in his earlier works was of special importance here. He became the expert in the theory of neutron scattering and in the theory of the slowing down of neutrons in matter.

When World War II broke out, Placzek naturally turned to the problems of the neutron propagation in nuclear chain reactions. He developed the most powerful methods for the treatment of the slowing down of neutrons by collision in matter and for the treatment of the diffusion of slow neutrons in matter.

The years after the war were devoted to further refinements and new developments in the theory of neutron scattering. This problem regained interest recently when the fundamental question of the electron-neutron interaction was raised. In order to identify the part of the scattering of neutrons in matter that is caused by this interaction, the theory of the scattering, elastic and inelastic, in solids and liquids had to be developed in all detail, and this was done by Placzek in his typically thorough and elegant way. Only with the help of his theories is it possible to interpret neutron scattering in crystals and liquids.

Unfortunately, Placzek did not write many papers. The style of his papers is impressive to the initiated in its elegance and conciseness, but it is hard reading for the outsider. The same was true of the few lectures that he delivered. This is why too few people know the importance of Placzek's contribution to physics and the great loss his death means for theoretical physics. The style of his thinking was true to the fine tradition of the classical period of theoretical physics, as exemplified by

Rayleigh, Lorentz, and his old friend, Kramers.[36] It is unfortunate that his long-suffered disease and his early death did not allow him many contacts with the younger generation, who do not always appreciate the value of style.

To his friends George Placzek was not only the expert in his field who was always ready to help and explain; he was a great support to them in times of stress and difficulty. They respected him for his clear sense of values and often went to him for counsel, knowing they would find a true helper with the vast experience of a full and interesting life. They have reason to mourn his passing.

*****

George Placzek
by Emilio Segrè [37]

George Placzek, permanent member of the Institute for Advanced Study, Princeton, N. J., died unexpectedly last October at the age of fifty in Zurich, Switzerland. Born in Brünn, Czechoslovakia, he studied at Vienna and later in the early thirties in Holland, France, Denmark, Italy, Germany, Israel, and Russia. During this period he mastered the languages and assimilated the best of the culture of the countries in which he resided and made friendships which were deep and close, lasting for life. His scientific work was mainly in the field of molecular physics and of neutron diffusion. It was characterized by a deep and thorough understanding of complicated phenomena. He simplified whenever he could, but never at the expense of completeness. His work on the Raman effect is a classic and dominates the field. The European turmoil of the 1930s brought him to the United States and during the war he worked at the Canadian atomic project and later at Los Alamos. After the war he spent some time at the General Electric Company, but he found his last task at the Institute for Advanced Study. Unfortunately his health became poor during the war and his last years were marred by physical disabilities endured with smiling stoicism. His published scientific work, although it does not reflect adequately his real contribution, is a permanent record of the scientist. It is hardly possible to give an adequate picture of the man. I believe that his friends

---

[36] Hendrik Anthony "Hans" Kramers (1894-1952) was a Dutch physicist who worked with Niels Bohr on electromagnetic waves interaction with matter.
[37] This obituary was published in *Physics Today*, 9(6), 41 (1956).

felt clearly his human greatness. Certainly he had an unerring judgment of human affairs, tempered by benevolence, and fortified by vast experience. The loss to science is great, the personal loss to his friends irreparable.

JOCULAR PHYSICS AT NIELS BOHR INSTITUTE

Chapter 6

Jocular Physics at Niels Bohr Institute

Excerpts from
Quantum Humor: The Playful Side of Physics at Bohr's
Institute for Theoretical Physics
by Paul Halpern
Phys. Perspect., 14, 279 (2012)

From the 1930s to the 1950s, a period of pivotal developments in quantum, nuclear, and particle physics, physicists at Niels Bohr's Institute for Theoretical Physics in Copenhagen took time off from their research to write humorous articles, letters, and other works. Best known is the Blegdamsvej Faust, performed in April 1932 at the close of one of the Institute's annual conferences. I also focus on the Journal of Jocular Physics, a humorous tribute to Bohr published on the occasions of his 50th, 60th, and 70th birthdays in 1935, 1945, and 1955. Contributors included Léon Rosenfeld, Victor Weisskopf, George Gamow, Oskar Klein, and Hendrik Casimir. I examine their contributions along with letters and other writings to show that they offer a window into some issues in physics at the time, such as the interpretation of complementarity and the nature of the neutrino, as well as the politics of the period. [...]

The Journal of Jocular Physics, Volume I

On October 7, 1935, the Institute celebrated Niels Bohr's 50th birthday. Rather than presenting a scholarly Festschrift, Bohr's current and former associates decided to prepare a satirical volume, which they dubbed the Journal of Jocular Physics. They reasoned that a serious Festschrift would make Bohr "feel it as his duty to read the contents and even try to learn something." Few copies of the original Journal of Jocular Physics have survived. Along with the script of the Blegdamsvej Faust, the Niels Bohr Archive reprinted its three volumes (1935, 1945, 1955) on the centenary of Bohr's birth in 1985. All share rich veins of humor and satire. The cover of Volume I (see Figure 6.1 on page 321) features a caricature by Piet Hein (1905-1996). It appears to be an image of Bohr, kneeling on the Earth, right hand on forehead, pondering deep thoughts, like Rodin's "The Thinker." Hein studied physics and mathematics at Bohr's Institute prior to the Second World War, and after the war became famous as an inventor, poet, artist, and designer.

The first article, entitled "Niels Bohr on His Fiftieth Birthday," was written in Japanese by physicist Kanetaka Ariyama and was translated into German. It is an ode to Bohr's excellent teaching and discusses progress in physics, including known particles, such as the neutron and positron, and the speculative neutrino, which, as in the Blegdamsvej Faust, plays a central and humorous role all three volumes.

One might wonder why the neutrino was considered to be so funny. First, it was proposed by Pauli, a much-satirized (and admired) physicist. Thoughts of the neutrino would have brought to mind Pauli's acerbic personality, his reputation as "The Scourge of God," and his alleged knack for disrupting experiments, the "Pauli effect," among other humorous characteristics. Second, the neutrino was famously elusive; neutrinos were seen as "ghost particles," "phantoms" that could be mistaken for delusions, products of too much abstract thinking, of tired minds that could fashion apparitions.

The contribution of Léon Rosenfeld (1904-1974) is a poem entitled "La Plainte du Neutrino" ("The Neutrino's Complaint"); it is a parody of "Un Secret" by the French poet Félix Arvers (1806-1850) and delves into the neutrino's mystique. While Arvers's poem is about a secret, unrequited love, Rosenfeld's parody probes the neutrino's elusiveness. For example, the last line of the original, "'Quelle est donc cette femme?' et ne comprendra pas," ("'What then is this woman?' and not comprehending,") is replaced by "'Quelle est cette énergie?' et ne comprendra pas," ("'What is this energy?' and not comprehending,") thus equating the mysterious woman with ghostly energy.

Japanese and French are only two of the languages represented: of the other nine papers, six are in German, two are in Danish, and one is in English. Weisskopf contributed two papers in German, one with Felix Bloch (1905-1983) as coauthor. The other papers were written by

JOURNAL
OF
JOCULAR PHYSICS

Niels Bohr Celebration Number
October 7, 1935

INSTITUTE OF THEORETICAL PHYSICS / COPENHAGEN

Figure 6.1   Title page of the first volume of the Journal of Jocular Physics (October 7, 1935). Credit: Reproduced by permission of the Niels Bohr Archive, Copenhagen.

Fritz Kalckar (1910-1938), Otto Robert Frisch (1904-1979) and George Placzek (1905-1955), Johan Ambrosen, Piet Hein, Hendrik Casimir, Hans Bethe (1906-2005), and Edward Teller (1908-2003). [...]

Weisskopf's single-author paper, entitled "Komplementäre Philosophie des Witzes" ("Complementarity Philosophy of Jokes"), explored the relationship between complementarity and humor. It had a serious slant; it asserted that while truth is difficult to find, jokes, through a version of complementarity, hold a kind of curved mirror up to reality. The result is a distorted view of truth that is nevertheless illuminating. Weisskopf's piece points to the enigmatic aspects of complementarity. This is perhaps the most abstract and perplexing of the key tenets of quantum mechanics. One common source of humor is a juxtaposition of opposites – the absurd aspects of contradictions. Complementarity and the uncertainty principle shatter the staid predictability of simple determinism, where in principle all properties of a system can be known simultaneously. Complementarity thus suits well the humor of the absurd; it is mentioned often throughout the three volumes.

Casimir's piece, entitled "Über eine weniger bekannt Bohr'sche Theorie und ihre experimentelle Bestätigung" ("On a less known Bohr Theory and its Experimental Verification"), included a poem about Bohr's reaction to American cowboy films like the Tom Mix movies. Casimir recalled that Bohr found cowboy movies implausible, and that he had a theory that predicted that the cowboy who draws his gun first is slower than the cowboy who then reacts and draws his gun. Casimir later translated his poem into English, which culminated in the lines:

> So the three of us went to the center of town
> And there at a gunshop spent many a crown
> On pistols and lead, and now Bohr had to prove
> That in fact the defendant is quickest to move.
> Bohr accepted the challenge without even a frown;
> He drew when we drew ... and shot each of us down.

*****

Preface to 1935 Bohr's Festschrift

At the occasion of Professor N. Bohr's fiftieth birthday it was originally planned to present him with a scientific "Festschrift." Since, however, there was the serious danger that Professor Bohr under such circumstances would feel it as his duty to read the contents and even to try to learn something from them and since it therefore did not seem advisable to impose upon him such an arduous task, it was instead decided to confine the scope of the proposed "Festschrift" to subjects of a Jocular character.

Such a decision appeared to be justified not only for exhilarating purposes; at the same time the great event to be celebrated could

```
              C O N T E N T S
              ________________


PREFACE

Niels Bohr zum fünfzigsten Geburtstag.  Kanetaka Ariyama
                        (mit deutscher Uebersetzung).

En Dag paa Bohrs Institut                F. Kalckar

Mitteilungen aus verschiedenen Gebieten.

Zur Frage der Messbarkeit des Landau'schen    O.R.Frisch und
            Schönheitskoeffizienten.          G.Placzek

Note on the Influence of Cosmetic Rays        J.Ambrosen
            on the Landau-Coefficient.

Zur Frage der Unterhaltung in                 F.Bloch und
            Physikergesellschaften.           V. Weisskopf.

Liv og Belæring i komplementerminologisk      Piet Hein
                        Belysning.

Ueber eine weniger bekannte Bohr'sche         H.Casimir
Theorie und ihre experimentelle Bestätigung.

Komplementäre Philosophie des Witzes.         V.Weisskopf

Die Wirkung mehrfach-periodischer             H.Bethe und
Bewegungen auf biologische Systeme.           E.Teller

La plainte du neutrino.                       L.Rosenfeld



     Interesting papers of high standard have been contributed
by G.Gamow, O.Klein and L.Rosenfeld.  Since, however, the
possibility of misinterpretation in a political and,therefore,
not purely jocular sense could not be entirely excluded, we
regret that they could not be published in the frame of this
volume.
```

Figure 6.2   1935 Niels Bohr Festschrift: Table of Contents.

properly be taken as an eminently appropriate occasion for a step, long since made an urgent need by the stormy development of our science, namely the organization and legalization of a comparatively old branch of physics: *The Jocular Physics* which for reasons hitherto not quite clearly understood has led a rather gloomy existence in the general frame of scientific efforts. As a first measure in this direction, the present volume inaugurates a new periodical, exclusively concerned with the Jocular aspect of the physical world.

It is however not the idea to open in this Journal a new playground for only too well-known tendencies which most adequately may be described as infantilism and are deplorably common among mature scientists. We are well aware that certain viewpoints of this kind have not been quite avoided in some of the minor contributions to this first volume; nevertheless it will not escape from the benevolent reader that the general background tries to preserve the very attitude of hopeful pessimism and serene preparedness which has shown to be so fruitful in other domains of atomic physics.

On the Question of Measurability of Landau's Beauty Coefficients[1]

O.R. Frisch and G. Placzek

London

The largely uniform distribution of women's beauty over the cities and countries of the globe has already been noted by early observers; however, all remarks thereon do not go beyond the framework of purely qualitative statements.

Landau has recently made a most striking attempt to quantitatively address the problem. This author suggested marking a city by the so-called beauty coefficients S defined by the equation

$$S = \frac{\sum_{j=1}^{3} n_j}{\sum_{j=1}^{5} n_j}$$

where n_1 to n_5 denote the numbers of the inhabitants (with exclusion of children and old people) of the corresponding city, which belong to the five Landau beauty classes.

For the determination of S, Landau created a statistical method, based on the assumption that the investigation of the modulation in question regarding the quantities n_j can be done by counting a traveling partial collective, for example of the passers-by at an imposing square. According to this method, counts for the studies of S were undertaken by Landau and collaborators in the cities Kharkov, Moscow, Leningrad, Turku, Stockholm, Copenhagen, i.a. ['among others']. These measurements led Landau to the establishment of far-reaching axiomatic statements in the area of the geo-aesthetic statistics.

However, doubts emerged about the adequacy of Landau's statistical measurement procedure due to counting attempts, which were undertaken [...] by us in collaboration with Farkas[2] in Tel Aviv in January 1935 for purely practical purposes. Namely, here the value $S = 0$ emerged, which however is completely contradictory to later experiences of both observers.

Due to this discrepancy it is reasonable to assume that the values obtained in the statistical method slightly differ depending on the choice of the partial collective. For the examination of this assump-

---

[1] Credit: Niels Bohr Archive, Copenhagen. Translated from German by Lena Funcke.

[2] Presumably, Ladislaus Farkas (1904-1948). L. Farkas was professor of physical chemistry at the Hebrew University. He died as a result of an air crash while on his way to the US.

tion, several measurements were undertaken under as varying external circumstances as possible, which will be reported in what follows.

Methods: At each step, 100 individuals were counted and simultaneously the class 1 to 3 numbers among those 100 were determined. For the purpose of impeccable counting, the traffic density has to lie within certain limits; in case of too little density, fatigue of the observer can easily occur, while in case of too large density, the resolution capacity can be exceeded.

A density of about 5 to 15 individuals (per minute) to be counted proved to be suitable. Very important is the exact identification of the sample collective; a line was constantly drawn in observer's mind from the viewpoint of the observer to the sidewalk. Only the individuals crossing this line were counted. In case of disregard of this rule, slightly higher values for S were obtained, due to the larger conspicuousness of the persons belonging to the classes 1 to 3.

In order to exclude any unintended influence of the observer on the counting, a method was elaborated, according to which the observer gets to know the counting result only after finishing the counting. For this purpose, the observer was equipped with a number of matches and put one of them into his vest pocket every time when a person belonging to the classes 1 to 3 was registered. After counting 100 individuals, the vest pocket was carefully emptied and its content was examined. In most cases both authors worked together. A few times, one and the same collective was examined by both authors independently; whereby the obtained S values always exactly coincided, even though the assignment of the single individuals to different classes was made very differently by both observers.

Results: [After numerous orientating preliminary tests, which served to develop the technique described above, we proceeded to performing the final measurements]. The enormous material, obtained through the final measurements is summarized in the following table:

Day	Hour	Place	S (in %)
Sept 7	8pm	Marble Arch	8
Sept 8	1 pm	Hyde Park	2
Sept 8	2 pm	Marble Arch	2
Sept 8	2-10 pm	Marble Arch	3
Sept 8	8 pm	Marble Arch	9
Sept 9	12-30 am	Picadilly	18
Sept 9	10 am	Picadilly	7
Sept 13	4 pm	Caledonian Market	0
Sept 13	4-15 pm	Caledonian Market	1
Sept 13	4-30 pm	Caledonian Market	0

As one can immediately see, the measurement results spread more strongly than it would be expected on purely statistical reasons. A precise analysis also shows that our guess of influence of time and the exact location of measurement was widely confirmed.

One could try to justify the detected differences. For example, the low values of the first two measurements from September 8 are apparently linked to the fact that on nice Sundays (September 8 was such a Sunday) a considerable part of the population in question – and indeed primarily the individuals belonging to the classes 1 to 3 – leave the city in the morning (the so-called "pair formation"). The especially low value from the Caledonian Market, on the other hand, could be traced back to the fact that mainly those members of social strata circulate there who currently lack both the knowledge and means for proper beauty care.

Anyway, apparently it follows from our measurements with certainty that Landau's statistical measurement method, also in its widely refined version created by us, does not allow determination of the beauty coefficient. On the basis of our experiences so far we would not exclude the possibility that further refinements, which would theoretically allow to quantitatively incorporate the influence of the above-mentioned selective factors – such as pair creation etc. – could actually allow the measurement of S. However, we must be prepared for an option

that the problem in question lies considerably deeper in the sense that
we could face here a complementarity situation, according to which
the fundamental measurability of S would have to be questioned. This
seems to follow from the fact that, in the considered case, it in principle
should not be possible to a priori exclude an uncontrollable interaction
between the measurement object and the observer with certainty.

One of us (O.R.F.) thanks Prof. Niels Bohr for granting of a travel
bursary, which enabled the implementation of the current study.

Literature:

[1] Plinius, *Hist. Nat. Cap*, 13, 7; Rabelais, *Gargantua und Panta-
gruel*, Vols. 1 and 2, Kap. 23; Further literature in e.g. Reitzenstein,
The Woman ['Weib' is an old version of 'Frau', which is the current
common German word for 'woman' – nowadays, 'Weib' is considered
as a derogatory term -LF] *in the Indigenous and Civilized Peoples.*

[2] Landau, Ukrainian notes and other publications.

Commentary -MS: The reference [2] above is fictitious. The very fact
that Landau had a "theoretical" classification of women is confirmed by
many sources. For instance, in the letter from Genia Kannegieser to
Rudolf Peierls (Sabine Lee, Volume 1, pp. 274 and 275) of June 2, 1931,
she wrote: "[S]end me as soon as possible the picture of Bretscher's
bride: she is the only lady that Dau classifies as the 1st class!"

In the letter of July 31, 1931, on pp. 336 and 337 Genia noted:

> Another sensational thing is the "situation" of Dau
> with a female who belongs to the 92nd class, according
> to Abbot [Matvei Bronstein]. Dau is of opinion she is
> the 2nd class. In any case she is rather vulgar-looking,
> with brightly rouged lips and yellow face because of freck-
> les. We weren't able to reveal her other features. She is
> married, and Dau is inclined to consider this situation
> asymmetric or zero. The husband is a terrible bore and
> Dau already became a pain in the neck to him, so the hus-
> band hates to see him. Dau and Abbot nearly drowned
> the innocent lady in the Neva River. In general, Dau
> tries his best. That is amusing, but he looks disgusting,
> is depressed and the "situation" (though Dau rejects this
> term) isn't jolly at all.

Figure 6.3   From the TV-documentary *Traum von Olympia – Die Nazi-Spiele von 1936.*
Placzek's last visit to Berlin in 1936 (tentatively).

$$*****$$

The following story strictly speaking is not directly related to George
Placzck. It gives another example of joyful mood and creative atmosphere which is so characteristic for young creative physicists at the
time when physics was on the rise. It refers to the Niels Bohr Institute
in the early 1930s when Placzek was there. At that time the Niels Bohr
group, along with the Göttingen group, were two main centers where
quantum nuclear physics was in the making.

Out of This World, by Max Delbrück[3]

In early summer there was a meeting in Zurich, followed for a few
of us by a few days in Ascona (Lago Maggiore). We swam, lay on
the beach and talked, and Gamow took a few snapshots. Back in

---

[3]Published in *Cosmology, Fusion and Other Matters*, Ed. F. Reines, (Colorado Associated University Press, Boulder, Co., 1972.)

Copenhagen one photograph intrigued him to no end and gave rise to a whole week's distracting discussions and experiments and to a little paper by Gamow and Rosenfeld, entitled "On the Determination of the Velocity of an Object Moving in a Fluid on the Basis of a Single Photograph."

Figure 1 of this paper shows Pauli in a bathing suit in the lake, gingerly stepping, his legs surrounded by surface ripples.

Here is the paper, translated from German and published for the first time after lying forty years in the archive of the Bohr Institute.

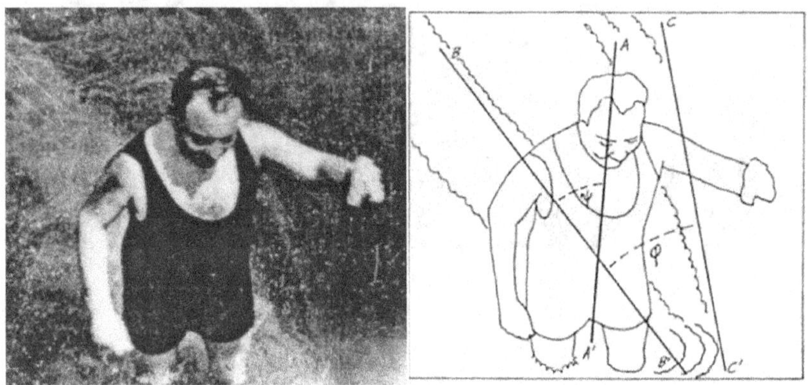

Figure 1

On the Determination of the Velocity of an Object Moving in a Fluid on the Basis of a Single Photograph

G. Gamow and L. Rosenfeld

It is well known that the motion of a solid object through a fluid produces surface waves which propagate with the group velocity (cf. Lamb, 1940)

$$w = \frac{1}{2}\,\frac{\lambda g}{2\pi} \tag{6.1}$$

(λ = wavelength, $g = 981$ cm sec^{-2}). Knowing the angle χ between the direction of motion of the object and the wave front, one obtains for the velocity of the object the simple relation

$$\frac{v}{w} = \operatorname{cosec}\chi. \tag{6.2}$$

In Figure 1 such wave groups can be discerned very clearly; but because of the unknown inclination of the photographic plate relative to the water surface one can not immediately read off the angle χ. It appears, however, that by measuring finer details of the picture this inclination α can be determined: it turns out that there are, spreading from the left leg of the object, circular waves which on the photograph appear as ellipses. From the axial ratios of these ellipses we obtain with sufficient accuracy $\cos\alpha = \frac{5}{11}$. We know further the projection ϕ of the unknown angle χ onto the photographic plate, as well as the angle ψ between the projections BB', AA' of the direction of motion and the vertical.

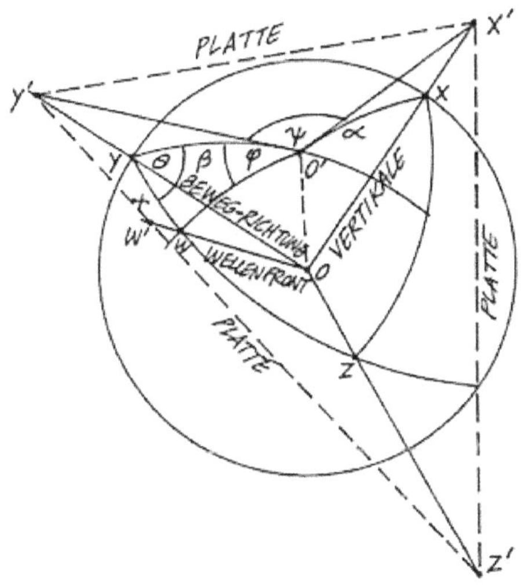

Figure 2

These data suffice to obtain χ with the help of a simple trigonometric argument which should be obvious without further explanation by inspection of Figure 2. We thus obtain

$$\mathrm{cotan}\,\chi = \frac{1}{\sin\beta}\left(\cos\beta\cos\theta + \sin\theta\,\mathrm{cotan}\,\phi\right). \qquad (6.3)$$

Here the auxiliary quantities β and θ are given by

$$\cos\theta = \sin\alpha\sin\psi, \quad \sin\beta = \cos\beta\,\mathrm{cosec}\,\theta. \qquad (6.4)$$

From the photograph we obtain $\phi = 32°40'$, $\psi = 48°42'$; thus from equation (3), $\mathrm{cosec}\,\chi = 2.2$.

We now obtain the wavelength λ from the photograph in connection with the known absolute size of the object. Thus $\lambda = 7\,\mathrm{cm}$. Equation

(1) now yields $w = 16.5$ cm sec^{-1} and finally equation (2) yields for the unknown velocity of the object $v = 36.3$ cm sec^{-1}.

We thank Professor Pauli cordially for suggesting this paper and our dear friend and colleague, Dr. Delbrück, for many critical remarks.

Copenhagen, Institut for teoretisk Fysik
June 7, 1931

***

The thanks to me at the end of this note were caused by the fact that I had told the authors a dozen times to leave me alone with this nonsense.

Bohr loved this paper and so did Ehrenfest, who came to visit. The latter submitted it to *Physica*, a Dutch physics teacher's journal. Alas, they did not accept it. Here is Ehrenfest's letter, returning the manuscript to Gamow and Rosenfeld:

Leiden, Holland
July 8, 1931

My dear colleagues:

I have the honor to inform you that Professor Fokker has presented your paper to the Editorial Committee of "Physica" and that your method was considered beautiful and interesting and that they, therefore, would be very happy to publish your paper provided that it is demonstrated with a more suitable object. I presume, if I understand correctly, that this means that it should be an object from nonliving nature or perhaps a plant or an animal.

Because of the shortness of the reply received by me I am not in a position to judge whether the Committee was afraid that there might be involved in the example chosen by you an incalculable correction, due to the influence of the so-called "soul" of the moving object.

Possibly the Committee was also influenced by a linguistic difficulty which one senses immediately if one tries in a logically precise way to define in any modern language the position of point A' which is so crucial for the whole method. I remain, my dear and respected contemporaries, as ever at your service,

Yours,
P. Ehrenfest

APPENDIX

Appendix

Fanny Ella von Halban Placzek

This brief appendix is devoted to George Placzek's wife, Fanny Ella Andriesse. She played a very special role in Placzek's life for twelve years, sharing with him all hardships and adventures of his life as a pioneer and trailblazer.

Fanny Ella Andriesse (everybody called her Els and that's how she called herself) was born on March 24, 1912 (see pages 44-45) into a wealthy Jewish family of Albert Andriesse and Ella Andriesse-van den Bergh. She had a brother Benno and a sister Rita. At the age of 21 she married a well-known French physicist (of Austrian-Jewish descent) Hans Heinrich von Halban (1908-1964), and in 1939 gave birth to a daughter Catherine Maulde, known as Maulde or Mauldely. Somewhat later she divorced Prof. von Halban and in 1943 married George Placzek (in the US).[1]

After Placzek's death in 1955, Els Placzek and George Placzek's stepdaughter lived in Switzerland. Later Maulde married James Griffin (b. 1933), who was a Fellow and Tutor in Philosophy at Keble College (Oxford) and then White's Professor of Moral Philosophy. They settled in Oxford. Els was a frequent guest in Oxford at least till 1993, when Maulde Griffin died. This is seen from Rudolf Peierls' diary (unpublished):

July 6, 1987

Next day I had dinner at the Griffins (Maulde Halban) –
a large and elegant dinner party with many philosophers,
some quite amusing, and gallons of champagne.

---

[1] Till recently, very little was known to the public about Els Placzek. This is in sharp contradistinction with the wives of other pioneers of nuclear physics. Literally a couple of days before sending this book to print we were lucky to obtain the recollections written by Ella (Elisabeth) Andriesse, Els Placzek's niece. The reader will find this essay on page 337. Thank you very much, Ella!

April 10, 1988

Last week I fed dinner to the Griffins and Els Placzek, who was visiting. Els is as ever charmingly naive and self-centered, though the charm is beginning to wear off a little.

December 31, 1989

Christmas dinner: again with venison: was only for the four of us. The day before I had run into Els Placzek, who was visiting the Griffins for a few days before taking them to Egypt. So we asked the three of them over for drinks.

March 9, 1992

Yesterday I called on the Griffins, where Els Placzek was again visiting. She is in very good form, as ever.

January 1, 1993

Monday night Nina [Byers] was out, and I invited Els Placzek to dinner, as she was leaving two days later. That was quite successful.

August 8, 1993

That weekend Maulde Griffin (Halban) died [on July 11] after a week of terrible agony. Evidently her cancer had spread to some other organ. Her mother, Els, was devastated. The funeral service was in Keble chapel (her husband is a fellow of Keble), the interment was for close family only.

Els Placzek lived a long life and died in 2014 at the age of 101. In 2005, Jan Fischer invited her as guest of honor to the International Symposium "Placzek 100" which was held in Brno, Czech Republic, on September 21-24. Unfortunately, she could not come for health reasons.

✶✶✶✶✶

Els Placzek, My Aunt
by Ella Andriesse

Fanny Ella (Els) Andriesse was born in Amsterdam, the Netherlands on March 24, 1912 and grew up in a warm, tight knit, non-observant Jewish family. She was the middle child with an older brother, Benno, and a younger sister, Rita.

Els' father was Albert Andriesse, a well known Dutch banker with a sharp mind and a keen sense of humor, to whom Els was very close. Her mother was Ella Andriesse-van den Bergh, the eldest daughter of Betsy and Samuel van den Bergh, a successful industrialist who was to become one of the principal founders of Unilever Co.

Els' grandparents played a big part in her life and from an early age, she would spend the summer holidays at their large home in Wassenaar[2] near the beach, together with her brother and sister and numerous cousins. There they played tennis together, went on long hikes and had lively conversations around the dinner table. Grandfather Samuel was active in Dutch politics and international Jewish causes and hosted guests from all walks of life, both Dutch and foreign. Els enjoyed these encounters, she was sociable and had enormous charm. Social responsibility and compassion for people in need also formed part of her upbringing.

Unlike her brother and sister, she did not go to university, but spent a year in Switzerland after high school to improve her French. Then she returned to Amsterdam to work as an assistant teacher at a kindergarten, a job she liked well, as she had a special affinity for children.

Not long thereafter she met her future husband, Hans von Halban, an Austrian-Jewish physicist on the ski slopes in the Swiss Alps. They were married in 1933, when she was 21, and moved together to Paris, where Hans worked. Els would never return to Holland to live. In 1939 their daughter Catherine Maulde (Mauldely) was born in France. Hans worked in Paris up until World War II, when he had to flee France with his young family while taking along the heavy water to England. The story goes that baby Mauldely sat on the heavy water during the journey (*Brighter than a thousand suns* by Robert Jungk). They lived the next years in Cambridge, UK.

---

[2]A municipality and town located in the province of South Holland, on the western coast of the Netherlands.

Figure A.1   Els as a teenager. Courtesy of Ella Andriesse.

In 1941 Els' parents, brother, sister and brother-in-law had managed to leave Nazi occupied Holland and had reached New York via Portugal. They settled in the US and eventually became US citizens, except Els' brother Benno. He re-enlisted with the Dutch army in Canada and spent the war years with a Dutch unit in Ceylon. He was the only immediate family member to return to Amsterdam after the war and he resumed his life there. He married in 1947 and had two children.

Figure A.2 Els with her grandfather Samuel van den Bergh at Portofino, Kulm Hotel, Italy. Courtesy of Ella Andriesse.

In 1942/43 the marriage of Hans von Halban and Els fell apart. Els married his colleague, George Placzek, and Hans also remarried. When

Els and George moved to Canada, Mauldely went to boarding school in the US.

I know very little about the life of Els and George together, except that they had a circle of loyal, close friends. With her good taste and generous personality, Els hosted many dinner parties and was a well-liked presence in the otherwise serious community of the scientists. She was fluent in several languages and felt at ease with different nationalities. Moreover, she had a wonderful sense of humor and she loved to laugh, not in the last place at herself. Whether in Montreal, Los Alamos, Princeton or Schenectady, Els and George forged lifelong friendships with his colleagues and their wives.

The loss of almost his entire family in the Holocaust haunted George after WWII and he suffered from recurrent depressions. Much too young, at age 50, his own life came to a sudden end, while Els was away in the US to be with her father after the death of her mother in 1955.

I never met George and I was 6 when he passed away. After his death Els settled in Switzerland, where she lived until her death in 2014. She became an avid mountain climber and a good skier. She enjoyed traveling to distant destinations.

Her annual visits to Holland to celebrate the St. Nicholas holiday with our family in December were highlights. Her elegant and joyful appearance lit up our home. Els was a very good listener and she was enthusiastically interested in whatever my brother and I were doing. As teenagers we spent lovely holidays with her in the Swiss Alps where we learned to ski and climb mountains. Her life's philosophy was to give people good memories, as those can never be taken away.

When one day I told her about a Dutch doctor in my ski class, who was working at Albert Schweitzer's hospital in Lambaréné in West Africa, she asked me to invite him over for drinks and she threw an animated cocktail party for him and some 20 of her friends. She had a warm personality and a generous spirit.

In the 1960s her daughter Mauldely and James Griffin had two children, Nicholas and Jessica, whom she loved dearly. Until well in her eighties Els could be seen skiing on the Swiss slopes with her offspring in tow.

Somewhere in mid-life Els began taking painting lessons in Paris and discovered that she had a real talent. For many years she spent part of the year in Paris to paint impressionist style paintings in warm colors.

Life did not spare Els sadness and grief. Ten years after the untimely death of George, she lost her dear father in 1965. In 1993 her beloved only daughter Mauldely died of cancer at age 53. Els struggled to come to terms with her loss and moved to an assisted living complex "La Gracieuse"[3] in Lonay near Lake Geneva. Yet, characteristically, she clung to her zest for life and she refused to surrender to despair. Only her painting stopped, as she had lost all inspiration. Her grandchildren became her main focus in life and eventually their children.

Els lived the last eighteen years of her life in "La Gracieuse," where she was happy and well settled. When she died in 2014 she was almost 102 years old.

Figure A.3 Els Placzek on February 20, 2005, in "La Gracieuse," Switzerland. Credit: From Jan Fischer private archive.

*****

---

[3] An assisted living complex for fairly independent older people.

1956 Zurich Police Report


```
                                                              27

                                          25.Januar 1956.
      27

Unters.Nr. 16420/55


                         An die Bezirksanwaltschaft, Büro 7,
                         Zürich.

                  Am 11.Oktober 1955 erteilten Sie uns mündlich (via
         Dr.med.J.Colombo) den Auftrag, an der Leiche des
                P l a c z e k , Georg, geb.26.September 1905, USA-Bürger,
                                  Dr.Physiker, wohnhaft gew.105 Battle
                                  Rd. Circle, Princeton, N.J./USA, tot
                                  aufgefunden am 10.Okt.1955 im Hotel
                                  "Im Park", Kappelistr.41, Zürich 2,
         eine gerichtliche Sektion sowie chemische Untersuchungen auf Gifte,
         insbesondere Schlafmittel, vorzunehmen und Ihnen über unsere Be-
         funde ein übliches Gutachten abzugeben.
                  Im vorliegenden Gutachten stützen wir uns auf:
         1. Sektionsprotokoll inklusive histologische Untersuchungen
            und pathologisch-anatomische Diagnose;
         2. chemische Untersuchung des Leichenmaterials auf Schlaf-
            mittel;
         3. Polizeirapport, dessen Kenntnis wir voraussetzen;
         4. verschiedene Rezepte und ärztliche Zeugnisse, sowie ver-
            schiedene Medikamentenpackungen, die uns von der Polizei über-
            geben wurden (wir werden auf diese Gegenstände, soweit sie
            für die Beurteilung des Falles eine Rolle spielen, eingangs
            des Gutachtens kurz eingehen).


                  (Die chemischen Untersuchungen waren recht zeit-
         raubend und wurden erst vor wenigen Tagen abgeschlossen, sodass
         sich die Abgabe des vorliegenden Gutachtens verzögert hat.)
```

Figure A.4   The first page of the Zurich Police Report of January 25, 1956. Courtesy of Ernest Kopp. For translation see pages 343 and 344.

Translation of the cover page of the Zurich Police Report of January 25, 1956 [4]

$\mathcal{N}^{\underline{0}}$ 16420/55

To the District Attorney's office, Bureau 7, Zurich

On the 11th of October 1955, (through Dr. Med. J. Colombo). Ordered examination of the corpse of

> *Placzek, Georg, born September 26, 1905,*
> *United States citizen, a physicist.*
> *Residence at the Battlefield Circle, Princeton, NJ, USA,*
> *died on 10th of October, 1955, in the hotel "Im Park,"*
> *Kappelistrasse 41, Zurich 2.*

Legal autopsy as well as chemical examinations on poisons, in particular sleeping substances, for our usual report.

In this report, we rely on:

1. Inspection protocol, including histological examinations and pathological-anatomical diagnosis;

2. Chemical examination of the corpse material for sleeping pills;

3. Police station data, of which we have preliminary knowledge;

4. Different prescriptions and medical certificates, as well as different medication packs, which were handed over to us by the police (we will discuss those items briefly in the beginning of the report to the extent they play a role for the assessment of the case).

(The chemical investigations were quite time-consuming and were concluded only a few days ago, so the delivery of the present report has been delayed.)

---

[4] Courtesy of Prof. Ernest Kopp. For the German original see page 342.

The concluding section of the Zurich Police Report of January 25, 1956[5]

On Monday the 10th of October 1955 the 50-year-old physicist Georg Placzek was found dead in his room at the hotel "Im Park". Dr. J. Colombo who pronounced him dead at about 21:00, estimated on the basis of the corpse changes that death should have occurred about 24 hours before. Dr. Placzek had been seen alive for the last time on Sunday evening 9th of October around 7:30 pm.

According to the police investigation Dr. Placzek had been previously treated by a number of physicians, he had already exhibited suicidal thoughts and had to be admitted in closed psychiatric facilities. His bedroom was a mess, which indicated that Dr. Placzek had took his clothes in a hurry. On the table there were several bottles of sleeping pills, including among others Somnifen (two original bottles, empty), a likewise empty pack of Nembutal tablets, various partially consumed partially intact packs of sleeping pills. Among the documents of Dr. Placzek there were found several medical certificates verifying that he had a sympathectomy (surgical removal of vegetative nerves on the both sides of the abdominal aorta) about five years before due to essential hypertension (increase of blood pressure). The operation had been successful and in the last few years Dr. Placzek has had more or less normal blood pressure under common blood pressure drugs. Among the medical prescriptions there were a lot of those for sleeping pills, especially for Somnifen, for Medinal, for Evipan, for Plexonal forte, all dated from June to August 1955.

The medical legal autopsy showed evidence of what is usually found in acute to subacute poisoning from sleep preparations, namely, strong general acute blood congestion, fairly strong cerebral edema, and agonal lung edema with strong hypostasis in both lungs. At the autopsy an intense smell of anise was already present in the stomach, and the gastric content contained whitish powder residues (apparently not fully dissolved and degraded pills). Anise is added to the liquid sleep preparation Somnifen as a flavor enhancer. The autopsy alone let it to assume that Dr. Placzek had taken a large quantity of Somnifen, and probably also of other sleeping pills. Along with these subacute processes the autopsy has also revealed a number of pre-existing pathological changes. Consistent with the documented sympathectomy five years earlier there were large old surgical scars on the both sides of the lower

---

back to be found. The heart was considerably enlarged which corresponds to meaningful cardiac hypertrophy. The weight of the heart was about 485 grams which is about 130 grams more than the average heart weight of a healthy male of the age of the deceased. This cardiac hypertrophy may be caused first of all by at least pre-existing hypertension. It is however quite possible that this hypertension reoccurred recently. The small arteries of the kidneys especially the so-called arterioles, exhibited a fairly high degree of thickening, that is, changes which are often found in the more severe forms of hypertension. Moreover, in the left kidney, there were sclerotic changes from old inflammatory processes, which however are of no great importance in regard to the cause of death. More important is that the cardiac muscle exhibited certain degenerative processes, namely, marked increase of the interstitial connective tissue, as well as multiple minor scars, i.e. sites where the muscle tissue was replaced by connective tissue.

These degenerative processes prove that the heart muscle was no longer properly supplied because of an existing rather strong atherosclerosis (which in particular impacts smaller branches), so that small areas were gradually depleted and replaced by connective tissue.

The autopsy presumptive diagnosis of poisoning with sleep preparations has been validated by chemical tests. In the stomach considerable quantities of Somnifen compounds as well as Pentobarbital (?) (which corresponds to Nembutal) could be isolated. It is interesting that in the liver and in the blood Somnifen (Numal) components but no Nembutal components could be found.

Somnifen is a fluid and its absorption occurs more rapidly than that of the Nembutal tablets which were used in this case.

The fact that it was Somnifen that was absorbed and detected in the liver as well as in the blood while Nembutal could only be detected in the stomach, does not necessarily indicate that Somnifen had to be taken earlier than Nembutal; this difference can also be explained by an almost simultaneous administration of both drugs. It is however important that the period of time between the intake of the drugs and the death could hardly be long, probably maximum of a few hours. This also means that Dr. Placzek was seen still alive on Sunday evening 9th October around 19:30 and that the death might have already occurred around 21:00 of that same day.

So rapid scenario is quite extraordinary for poisoning with sleeping preparation. Only with overdose and rapid absorption of a sleeping agent can death due to poisoning with sleep preparation occur within a few hours. In the present case considerable amounts of sleeping agent were present in the stomach, but not in the organs. This suggests that poisoning with a sleeping preparation was not the only cause of death, but rather a more or less important component. The major cause of death may have been heart failure due to the previously discussed pre-existing heart changes. Experience has shown that in cardiac hypertrophies and cardiac muscle degeneration the main factor for the cause of death could be heart failure due to the previously discussed heart changes.

Experience has shown that in cardiac hypertrophies and cardiac muscle degeneration, which was the case of Dr. Placzek, cardiac paralysis can already occur spontaneously or for a minor reason.

Now, however, on the basis of the autopsy findings and the chemical tests it must be assumed that Dr. Placzek had taken some time before his death a considerable amount of sleeping agents. From the autopsy we can not estimate whether it was a fatal dose (without pre-existing heart disease), or only a more or less severe overdose. The prehistory perhaps speaks in favor of a severe overdose in suicidal intention. The possibility that Dr. Placzek (possibly under the effect of a certain sleeping agent) took an extra dose and thereby involuntarily (i.e., being in the state of reduced judgment) more or less overdosed, cannot be totally excluded.

In practice, this uncertainty whether ingestion of a large quantity of sleeping pills in suicidal intention or involuntarily occurred, plays no significant role. It is essential that a crime can be excluded. It is practically impossible to secretly administrate such a large quantity of sleeping pills to a man in the healthy state of mind; in particular, this applies to the liquid Somnifen, which has a disgusting taste due to the taste-enhancer (anise).

Summary.

1. The autopsy of the body of the 50-year-old Dr. Placzek revealed findings that speak for a subacute sleeping pill poisoning.

2. Chemical testing revealed considerable amounts of Somnifen and Nembutal in the gastric contents, as well as smaller amounts of Somnifen components in the liver and in the blood.

3. The autopsy revealed pre-existing pathological changes, first of all relatively severe cardiac hypertrophy with myocardial degeneration due to rather severe atherosclerosis.

4. The death occurred due to intoxication with sleeping agents in the presence of the pre-existing cardiac changes.

5. The autopsy alone does not allow to distinguish whether the intake of sleeping agents occurred with suicidal intention or more or less involuntarily.

6. There is no evidence of a crime; administration of sleeping agents with criminal intention may be excluded.

Br.raed. E. Hardmcier, Chief Physician

*****

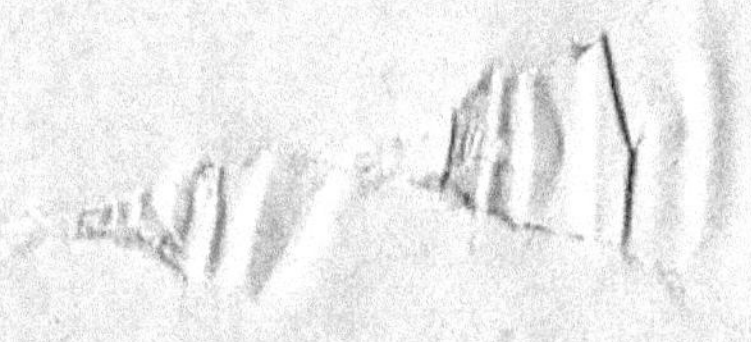

Figure A.5 The title page of Placzek's "Raman Effect" in Russian, Kharkov, 1935, translated from German by A.S. Kompaneyets and E.M. Lifshitz. Editor L. Rozenkevich.

Los Alamos National Laboratory (1945)

THEORETICAL DIVISION
MAY 10, 1945

H.A. Bethe, Division Leader, E-208, Ext 71
V.F. Weisskopf, Deputy Division Leader, E-217, Ext 74
J. von Neumann, Consultant, E-205 , Ext 72 R3

GROUP T-1
R. Peierls, E-119, Ext 178
R.F. Christy, Section Leader E-120, Ext 178 R2
K. Fuchs, Section Leader, E-118, Ext 77
Baroody, E.M., E-121, Ext 178 R2
Calkin, J.W., E-117, Ext 77
Inglis, D.R. T-35, Ext 54
Keller, J., E-120, Ext 178 R2
Penny, W.G., E-101A, Ext 470
Podgor, T/5 S., E-116, Ext 469
Roberts, T/5 A.E., E-117, Ext 77
Skyrme, T.H.R., E-118, Ext 77
Stark, R.H., E-116, Ext 469
Stein, T/5 P.R., E-121, Ext 178

GROUP T-2
R. Serber, Leader, E-109, Ext 177
L.I. Schiff, Alt. Leader, E-108, Ext 76
Case, K.M., E-107, Ext 76
Glauber, R., E-107, Ext 76
Kurath, T/3 D., E-202, Ext 205
Rarita, W. E-111, Ext 177
Richman, C., E-110, Ext 177
Stehle, T/5 P., E-107, Ext 76

GROUP T-3
V. Weisskopf, Leader, E-217, Ext 74
R.E. Marshak, Alt. Leader, E-218, Ext 74 R2
Bellman, Pvt. R., E-222, Ext 468
Cohen, T/5 S., E-219, Ext 74 R2
Lennox, T/5 E., E-218, Ext 74 R2
Olum, Paul, E-216, Ext 74
Smith, J.H., E-219, Ext 74
Wing, Milton, E-222, Ext 468
Bowers, W.A., E-216, Ext 74

GROUP T-4
R. Feynman, Leader, E-206, Ext 72
J. Ashkin, E-209, Ext 72 R2
Ehrlich, R., E-210, Ext. 72 R2
Peshkin, T/4 N., E-203, Ext. 79 R3
Reines, F., E-210, Ext 72 R2
Welton, T.A., E-209, Ext 72 R2

GROUP T-5
D. Flanders, Leader, E-205, Ext 69 R1
P. Whitman, Alt. Leader, E-204, Ext 69 R2
Atkins, A.L., E-211, Ext 73 R2
Davis, R.R., E-215, Ext 74
de le Vin, E., E-201, Ext 79
Elliott, J., E-213, Ext 73 R1
Hauser, T/5 F.H., E-214,
Huber, T/4 .C., E-120, Ext 70
Hudson, H., E-214, Ext 73 R1
Inglis, B., E-212, Ext 73 R2
Johnson, M., E-212, Ext 73 R2
Kellogg, T/5 H., E-203, Ext 69 R3
Langer, B., E-201, Ext 79
Page, T/3 W., E-214, Ext 73 R1
Rau, E., T/3, E-120, Ext 70
Staley, T/3 J., E-212, Ext 73 R2
Teller, M., E-214, Ext 73 R1
Vuletic, T/5 V., E-211. Ext 73 R2
Wilson, F., E-212, Ext 73 R2
Wright, T/5 E., E-213, Ext 73 R1
Young, T/Sgt., G., E-211, Ext 73 R2

GROUP T-6
E. Nelson, Leader, E-116
N. Metropolis, Alt. Leader, E-115, Ext 78
R. Feynman, Consultant, E-206, Ext 72
Ewing, F.E., E-112, Ext 75
Goldberg, T/5., E-108, Ext 76
Kemeny, Pvt. J., E-105, Ext 75
Hamming, R.W., E-114, Ext 78
Heermans, Corp. A., E-105, Ext 75
Heller, T/5 A., E-105, Ext 75
Hurwitz, T/5 D., E-105, Ext 75
Johnston, T/3 J., E-103, Ext 75
Kington, T/4 J., E-105, Ext 75
Livesay, N., E-112, Ext 75
Ninger, H., E-105, Ext 75
Noah, F.E., E-105, Ext 75
Vorwald, T/5 A., E-105, Ext 75
Zimmerman, T/3 W., E-105, Ext 75

GROUP T-7
J. Hirschfelder, Leader, T-30, Ext 206
J. Magee, Alt. Leader, T-30, Ext 206
Brummer, T/5 E., T-28, Ext 206
Feckete, T/4 P., T-26, Ext 206
Larson, T/4 L., T-28, Ext 206
Ostrow, E., T-30, Ext 206
Schwartz, T/4 P., T-28, Ext 206

GROUP T-8
G. Placzek, Leader, E-220, Ext 468
Mark, C., E-221, Ext 468 R2
Carlson, B., E-221, Ext 468 R2
Day

Figure A.6 Los Alamos Theory Division, May 10, 1945. George Placzek is the Leader of T-8. Courtesy of Cristian Batista and Los Alamos Bradbury Museum of Science.

200   N A T U R E   July 29, 1939, Vol. 144

LETTERS TO THE EDITORS

The Editors do not hold themselves responsible for opinions expressed by their correspondents. They cannot undertake to return, or to correspond with the writers of, rejected manuscripts intended for this or any other part of Nature. *No notice is taken of anonymous communications.*

Notes on points in some of this week's letters appear on p. 208.

Correspondents are invited to attach similar summaries to their communications.

Nuclear Reactions in the Continuous Energy Region

It is typical for nuclear reactions initiated by collisions or radiation that they may, to a large extent, be considered as taking place in two steps : the formation of a highly excited compound system and its subsequent disintegration or radiative transition to a less excited state. We denote by A, B, . . the possible alternative products of the reaction, specified by the nature, internal quantum state, and spin direction both of the emitted particle or photon and of the residual nucleus and the orbital momentum. Further, we call P_A, P_B . . . the probabilities, per unit time, of transitions to A, B, . . . respectively, from the compound state.

The cross-section for the reaction $A \rightarrow B$ is then evidently

$$\sigma_B^A = \sigma^A \frac{P_B}{P_A + P_B + \cdots}, \qquad (1)$$

where σ^A is the cross-section for a collision in which, starting from the state A, a compound nucleus is produced. This formula implies, of course, that we are dealing with energies for which the compound nucleus can actually exist, that is, that we are either in a region of continuous energy values or, if the levels are discrete, that we are at optimum resonance. Moreover, it is assumed that all possible reactions, including scattering, proceed by way of the compound state, neglecting, in particular, the influence of the so-called 'potential scattering', where the particle is deflected without actually getting into close interaction with the individual constituents of the original nucleus.

On these assumptions a very general conservation theorem of wave mechanics[1] yields the relation

$$\sigma^A = \frac{\lambda^2}{\pi}(2l+1) \frac{P_A}{P_A + P_B + \cdots}, \qquad (2)$$

where λ is the wave-length of the incident particle and l is the angular momentum.

In the case of discrete levels, (1) and (2) give the same cross-section as the usual dispersion formula, if one applies it to the centre of a resonance level and neglects the influence of all other levels. In this case we have for each resonance level a well-defined quantum state of the compound nucleus, and its properties, in particular the probabilities P_A, P_B, . . . then cannot depend on the kind of collision by which it has been formed, that is, they would be the same if we had started from the fragments B, or C, . . instead of A.

In the case of the continuum, however, where there are many quantum states with energies that are indistinguishable within the life-time of the compound nucleus, the actual state of the system is a superposition of several quantum states and its properties depend on their phase relations, and hence on the process by which the compound nucleus has been produced.

This dependence is made particularly obvious if we consider the formula

$$\overline{\sigma^A} = \frac{\hbar}{2} \rho \lambda^2 (2l + 1) P_A^0, \qquad (3)$$

for the mean value of the cross-section over an interval containing many levels, which follows from the well-known considerations of detailed balancing. Here ρ is the density per unit energy of levels (of suitable angular momentum and symmetry) of the compound nucleus. P_A^0 is the probability for process A in statistical equilibrium and thus refers to a micro-canonical ensemble of compound states built up from the fragments A, B . . . respectively, with proper statistical weights.

In the case of discrete levels, where formula (3) can also be derived directly from the dispersion formula, P_A^0 is simply an average over the individual levels of the probability P_A, which in this case is well defined.

In the continuum, (3) must be identical with (2), since the cross-section does not vary appreciably over an energy interval containing many levels, and hence, comparing (2) and (3)

$$\frac{P_A^{(A)}}{P_A^0} = \frac{\pi}{2} \hbar \rho \, (P_A^{(A)} + P_B^{(A)} + \cdots) = \frac{\pi}{2} \frac{\Gamma^{(A)}}{d}, \qquad (4)$$

where the superscript A has been added to the probabilities occurring in (1) in order to show explicitly the dependence on the mode of formation, and where $\Gamma^{(A)}$ is the total energy width of the compound state concerned and $d = \frac{1}{\rho}$ the average level distance. In the continuum, where $\Gamma^{(A)} \gg d$, the probability $P_A^{(A)}$ of re-emitting the incident particle without change of state of the nucleus will thus be much larger than the probability of the same process in a compound nucleus produced in other ways.

While the arguments used so far are of a very general character, more detailed considerations of the mechanism of nuclear excitation are required for a discussion of the dependence $P_B^{(A)}$ of the mode A of the compound nucleus provided $A = B$.

One can think of cases in which such a dependence must obviously be expected; in fact, if a large system be hit by a fast particle, the energy of excitation might be localized in the neighbourhood of the point of impact, and the escape of fast particles from this neighbourhood may be more probable than in statistical equilibrium. Further, if the system had modes of vibration very loosely coupled, the excitation of one of them, for example by radiation, would be unlikely to lead to the excitation of a state of vibration made up of very different normal modes, even though the state may be quite strongly represented in statistical equilibrium.

In actual nuclei, however, the motion cannot be described in terms of loosely coupled vibrations, nor

Figure A.7   N. Bohr, G. Peierls, and G. Placzek, *Nature*, Vol. 144, № 3639, 200 (1939).

would one expect localization of the excitation energy to be of importance in nuclear reactions of moderate energy. If we suppose that there are no other special circumstances which would lead to a dependence of $P_B^{(A)}$ on A, it is thus a reasonable idealization to assume that, even in the continuum, all $P_A^{(A)}$ are equal to P_B^0, except, of course, for $A = B$, where we have seen in (4) that the phases are necessarily such as to favour the re-emission of the incident particle.

A typical case of a reaction in the continuum is the nuclear photo-effect in heavy elements, produced by γ-rays of about 17 mv. In the first experiments of Bothe and Gentner, there seemed to be marked differences between the cross-sections of different elements, but the continuation of their investigations[2] indicated that these differences can be accounted for by the different radioactive properties of the residual nuclei, and that the cross-sections of all heavy nuclei for photo-effect are of the order of 5×10^{-26} cm.[2].

In previous discussions, based on formulæ (1) and (2), where the distinction between $P_A^{(A)}$ and P_A^0 was not clearly recognized, it was found difficult, however, to account for photo-effect cross-sections of this magnitude. In fact, if one estimates the probability of neutron escape P_B at about 10^{17} sec.$^{-1}$, one should have for P_A 10^{16} sec.$^{-1}$, and as long as this was taken as P_A^0 it seemed much too large, since it evidently must be much smaller than the total radiation probability, estimated at about 10^{15}, which included transitions to many more final levels besides the ground state.

We see now, however, that $P^{(A)}$ is here considerably larger than P_A^0, since the level distance at the high excitations concerned is probably of the order of 1 volt, whereas the level width corresponding to the above value of P_B is about 100 volts. From (4), or more directly from (3), P_A^0 is thus seen to be only about 10^{13} sec.$^{-1}$, which would appear quite reasonable.

N. BOHR.
R. PEIERLS.
G. PLACZEK.

Institute of Theoretical Physics,
    Copenhagen.
      July 4.

[1] The details of this and of the other arguments of this note will be published in the *Proceedings of the Copenhagen Academy*.
[2] Bothe, W., and Gentner, W., *Z. Phys.*, **106**, 236 (1937); **112**, 45 (1939).

The Scattering by Uranium Nuclei of Fast Neutrons and the Possible Neutron Emission Resulting from Fission

THE work to be described concerns only fast neutrons, and its object is the study of their scattering by uranium and the possible neutron emission which accompanies the fission of the nucleus.

The experiments were performed with a polonium plus beryllium source equivalent to 3mC. of radon plus beryllium. An ionization chamber surrounded with 2·5 cm. lead, filled with hydrogen at a pressure of 35 atm., was used as a neutron detector. The insulated electrode was connected to a compensated electrometer valve[1], the grid leak being 10^{11} ohms and the sensitivity $1·2 \times 10^{-15}$ amp./div. on the scale.

We have employed two experimental arrangements in which the source was placed (1) between the chamber and the substance used as scatterer, the nature and the thickness of which were variable; (2) in the centre of a cube of 16 cm. side, alternately filled with uranium oxide (specific gravity, $d = 4·0$) and lead oxide (compressed to $d = 3·8$).

The first type of experiment gave us the total scattering cross-section, which is, as can be shown, $\sigma_t = \sigma_e + k_i \sigma_i$; for uranium oxide $\sigma_t = \sigma_e + k_i \sigma_i + k_r v_r \sigma_r$, where σ_e, σ_i, σ_r are respectively the average cross-sections of elastic and inelastic scattering and of fission; v_r is the average number of neutrons produced per fission; k_i and k_r are the average efficiency factors of the chamber for the neutrons having undergone an inelastic collision or for the neutrons resulting from fission. · The efficiency for the direct neutrons was taken to be unity, $k = 1$. For neutrons elastically scattered by nuclei of sufficiently high mass, $k_e = k = 1$. We have calculated k, taking into account the size of the chamber, the cross-section for proton projection, etc. The spectrum of polonium plus beryllium neutrons has been considered[2] to contain 50 per cent of neutrons of W_n less than 10^5 ev. We thus obtain :

$10^{-6} W_n$	0·1	0·5	3	5	10	ev.
k	0·3	1	1·9	1·7	1·2	

In view of a possible extrapolation that would give $\sigma_e + k_i \sigma_i$ for uranium, we have in the same way experimented with scattering by lead oxide, lead, copper and zinc.

The results of the first experiment were as follows :

Substance	Cu	Zn	Pb	PbO$_2$	UO$_2$	(O)calc.	(U)calc.
$\sigma_t \times 10^{-24}$ cm.^{2} ($\pm$ 10%)	2·2	2·3	5·4	9·5	14·4	2	10·3

The values for uranium and oxygen are calculated on the assumption of the additivity of the cross-sections in lead oxide and uranium oxide.

The second experiment gives us, in the first approximation, the absorption coefficient $(1-k_i)\sigma_i + (1-k_r v_r)\sigma_r$, the value of σ_e being only as a correction term in the determination of the mean free path λ and the average distance L travelled by the neutrons before they escape from the whole mass, which is supposed spherical, the radius being r and $L = r\left(1 + \frac{1}{4}\frac{r}{\lambda}\right)$. This experiment, taking into account the results of the previous experiments, gives for lead, $(1-k_i)\sigma_i \simeq 2 \times 10^{-24}$ cm.^{2}. Assuming that σ_i can reach 30 per cent of σ_e[3], this gives $k_i (\simeq) 0$.

With the exception of uranium, for which one must consider not only σ_i, but also $v_r \sigma_r$, it is probable that σ_i is not very different from σ_e because of the small value of k_i.

In the case of uranium, however, we have,

$$(1-k_i)\sigma_i + (1-k_r v_r)\sigma_r \simeq 0·9 \times 10^{-24} \text{ cm.}^2 \qquad (1),$$

or, by adding to σ_t, thus eliminating k_i and k_r,

$$\sigma_e + \sigma_i + \sigma_r \simeq 11·2 \times 10^{-24} \text{ cm.}^2. \qquad (2).$$

If it is supposed that each fission produces radio-elements, the cross-section measured by Joliot, and by Anderson *et al.*[4] would be identical with σ_r, which they found to be $\sigma_r \simeq 10^{-24}$ cm.^{2}. In this case we see that $(\sigma_e + \sigma_i)$ is much greater ($\simeq 11·1 \times 10^{-24}$ cm.^{2}) than that given by an extrapolation ($\simeq 6 \times 10^{-24}$ cm.^{2}).

On the other hand, it results from (1) that, if the value of σ_i is comparable to that of the next elements (1 to 2×10^{-24} cm.^{2}), v_r can, with plausible assumptions as to the coefficients k_i and k_r, take variable values, for example, from 1 to 5, or even more.

One can see that, so long as σ_i is not determined separately, the experiments of the type described do not allow us to determine v_r and σ_r (characteristics of the fission), or to conclude that neutrons are

Figure A.8   Continuation of Fig. A.7.

PHYSICAL REVIEW   VOLUME 86, NUMBER 3   MAY 1, 1952

The Scattering of Neutrons by Systems of Heavy Nuclei*†

G. PLACZEK
Institute for Advanced Study, Princeton, New Jersey
(Received December 20, 1951)

It is shown that the scattering of neutrons by a system of heavy nuclei may, for neutron energies that are large compared with the level separation of the system, be described in terms of averages of simple two-particle operators over the initial state. Expressions are derived for the total and differential cross sections and for the first few moments of the energy transfer. The expression for the cross section can be explicitly evaluated even for complicated scattering systems and leads to an accurate representation of the energy dependence of the nuclear scattering. The results are applied to the problem of the detection of small electronic contributions to the cross section.

1. INTRODUCTION

WE consider the collision of a neutron of mass m, energy E_0, wave vector $\mathbf{k}_0$, and wave number k_0 with a system of N nuclei—such as a molecule, a gas, a liquid, or a crystal—in the state "a" of energy E_a. The standard Born approximation[1] leads to the scattering cross section,

$$\sigma^{(a)} = \frac{2}{k_0} \sum_b \int |(F(\mathbf{\kappa}))_a{}^b|^2 \delta \left\{ \kappa^2 - 2\mathbf{k}_0 \cdot \mathbf{\kappa} + \frac{2m}{\hbar^2}(E_b - E_a) \right\} d\mathbf{\kappa}. \quad (1.1)$$

The summation extends over all states of the nuclear system, and the integration variable $\mathbf{k}$ is the momentum transferred to the system, divided by $\hbar$. The delta-function expresses conservation of energy and the scattering amplitude $F(\mathbf{\kappa})$ is given by

$$F(\mathbf{\kappa}) = \sum_{s=1}^{N} a_s \exp(i\mathbf{\kappa} \cdot \mathbf{r}_s). \quad (1.2)$$

Here $\mathbf{r}_s$ is the position vector of nucleus s and a_s its scattering length.

The scattering lengths will be taken to be independent of E_0 and $\mathbf{\kappa}$ in the energy region considered, they may, however, depend on the nuclear spin variables. For nuclei without spin the scattering length a_s is simply related to the scattering cross section σ_s of the bound nucleus by

$$\sigma_s = 4\pi a_s{}^2, \quad (1.3)$$

and the matrix element of the scattering amplitude may here be written

$$F_a{}^b = \int \psi_b{}^*(\mathbf{r}_1, \cdots \mathbf{r}_N) F \psi_a(\mathbf{r}_1, \cdots \mathbf{r}_N) d\mathbf{r}_1 \cdots d\mathbf{r}_N, \quad (1.4)$$

where the ψ's are the eigenfunctions of the Hamiltonian H of the motion of the nuclei in the field of the interatomic forces.

For nuclei with spin, (1.3) and (1.4) have to be modified which will be taken care of in Sec. 7.

The straightforward evaluation of (1.1) requires detailed knowledge of the eigenfunctions of all accessible states. It is therefore possible for the very simplest cases only, and even there it becomes progressively more laborious with increasing neutron energy. In the following it will be shown, however, that for systems of heavy nuclei the cross section may be expressed in terms of averages of simple operators over the initial state as soon as the neutron energy is large compared to the level separation of the system.

2. GENERAL CONSIDERATIONS

In the limit of infinite neutron energy, (1.1) goes over into[2]

$$\sigma_\infty^{(a)} = \sum_s \left(1 + \frac{m}{M_s}\right)^{-2} \sigma_s, \quad (2.1)$$

where M_s is the mass of nucleus s. We shall have to determine, in particular, the behavior of the cross section on the approach to this limit. While certain general statements in regard to this problem may be made on the basis of the classical particle picture,[3] their application to the case of heavy nuclei would appear to be of limited practical use without a closer investigation. This may be seen as follows. In classical mechanics the collision—in our case of contact interaction—takes place instantaneously and the energy dependence of the cross section is determined by the momentum distribution of the scattering particles. The cross section of a system under the influence of forces will thus be equal to that of a free system with the same momentum distribution. A necessary condition for the description of the collision in classical particle terms requires that the collision time h/E_t, where E_t is the classical energy transfer, be short compared to the periods of the

---

* The results of this paper were presented at the International Conference on Nuclear Physics, Basel, September, 1949.
† Work partially supported by the AEC and ONR.
[1] For the justification of the use of the Born approximation in the treatment of this problem see E. Fermi, Ricerca sci. **7**, 13 (1936); G. Breit, Phys. Rev. **71**, 215 (1947).

[2] E. Fermi, Ricerca sci. **7**, 13 (1936).
[3] G. Placzek, Phys. Rev. **75**, 1295 (1949); and as quoted in reference 7.

system. Accordingly, the energy transfer $(m/M)E_0$ has to be large compared to the level separation of the system, and thus

$$E_0 \gg (M/m)\Delta, \qquad (2.2)$$

where Δ is the level separation. Indeed, one can see immediately that unless condition (2.2) is satisfied the energy distribution of the scattered neutrons must differ appreciably from the classical energy distribution.

For heavy nuclei (2.2) becomes extremely restrictive and we shall therefore attempt to study the collision under the much less restrictive condition,

$$E_0 \gg \Delta. \qquad (2.3)$$

For this purpose we start from the static approximation which results from neglecting the energy changes associated with the transition of the system. It is obtained by omitting the last term in the delta-function in (1.1). The summation over b can then be carried out immediately by closure so that (1.1) goes over into

$$\sigma^{(a)} = \frac{2}{k_0} \int (F^*(\kappa)F(\kappa))_a{}^a \delta(\kappa^2 - 2\mathbf{k}_0 \cdot \kappa)d\kappa$$

$$= \int (|F(\mathbf{k}_0 - k_0\boldsymbol{\Omega})|^2)_a{}^a d\Omega, \qquad (2.4)$$

where $\boldsymbol{\Omega}$ is a unit vector. Equation (2.4) represents the average of the cross section of a system, consisting of nuclei fixed in definite positions, over the configurations of the initial state of the real system. It may thus be referred to as the static approximation. This approximation forms the basis of the theory of x-ray scattering by molecules, crystals, and liquids.[4]

The cross section (2.4) has to be well distinguished from the elastic cross section (or more precisely, the cross section for scattering without change of quantum state) which from (1.1) is given by

$$\sigma_{el}{}^{(a)} = (2/k_0) \int |(F(\kappa))_a{}^a|^2 \delta(\kappa^2 - 2\mathbf{k}_0 \cdot \kappa)d\kappa$$

$$= \int |(F(\mathbf{k}_0 - k_0\boldsymbol{\Omega}))_a{}^a|^2 d\Omega. \qquad (2.5)$$

In the literature the cross section is frequently divided into an elastic and inelastic part and the energy dependence of each part is discussed separately. It is clear from (2.4) that in the static approximation this separation is an unnecessary complication which has to be avoided as far as possible. This situation will be shown later to hold also in higher approximations.

For systems consisting of one nucleus only, the static approximation leads to a constant cross section. Since

---

[4] See the review article by M. Born in Reports of Progress of Physics **9**, 294 (1943).

here $|F|^2 = a_s{}^2$, we have from (2.4)

$$\sigma^{(a)} = \sigma_s. \qquad (2.4a)$$

For systems consisting of several nuclei the evaluation of (2.4) involves the probability distribution of internuclear distances in the initial state (pair density). In the limit of infinite neutron energy the cross section (2.4) reduces here to

$$\sigma_\infty{}^{(a)} = \sum_s \sigma_s, \qquad (2.6)$$

which differs from the correct result (2.1) by the absence of the reduced mass factors.

It will now be our problem to establish the exact conditions of validity of the static approximation which neglects the energy changes and to improve it by taking account of the energy changes. We shall limit ourselves to systems of random orientation. This will contribute to brevity and the generalization of the results to oriented systems such as single crystals will become quite obvious in the course of the treatment.

For systems of random orientation we may average the delta-function in (1.1) over all directions of $\mathbf{k}_0$, whereby (1.1) goes over into

$$\sigma^{(a)} = \frac{1}{2k_0{}^2} \sum_b{}^{E_b < E_0 + E_a} \int_{a,b} |F_a{}^b(\kappa)|^2 \frac{d\kappa}{\kappa}, \qquad (2.7)$$

where the subscript a, b under the integral sign indicates that the integration extends over a region in κ-space bounded by the spheres of radius $k_0 + k_{ab}$ and $|k_0 - k_{ab}|$ with

$$k_{ab}{}^2 = 2m\hbar^{-2}(E_0 + E_a - E_b). \qquad (2.8)$$

Introducing the average $\phi_{ab}(\kappa^2)$ of the square of the matrix element over all directions of κ

$$\phi_{ab}(\kappa^2) = (1/4\pi) \int |F_a{}^b(\kappa)|^2 d\Omega_\kappa \qquad (2.9)$$

and putting

$$(E_b - E_a)/E_0 = x_{ab}, \qquad (2.10)$$

we may write (2.7)

$$\sigma^{(a)} = \frac{\pi}{k_0{}^2} \sum_b{}^{x_{ab} < 1} \int_{k_0{}^2\{1 - (1 - x_{ab})^{\frac{1}{2}}\}^2}^{k_0{}^2\{1 + (1 - x_{ab})^{\frac{1}{2}}\}^2} \phi_{ab}(\kappa^2)d\kappa^2. \qquad (2.11)$$

If in this expression the quantities x_{ab} are neglected throughout, which implies also the replacement of the restricted sum over b by an unrestricted one, it goes over into

$$\sigma^{(a)} = \frac{\pi}{k_0{}^2} \int_0^{4k_0{}^2} \left(\sum_b \phi_{ab}(\kappa^2)\right)d\kappa^2$$

$$= \frac{1}{4k_0{}^0} \int_0^{4k_0{}^2} d\kappa^2 \int_0 (|F(\kappa)|^2)_a{}^a d\Omega_\kappa, \qquad (2.12)$$

which is the result of the static approximation (2.7), averaged over all directions of $\mathbf{k}_0$.

The expression (2.11) has a form suitable for taking the energy exchanges into account. For heavy nuclei, owing to the difficulty of large energy transfers between light and heavy particles, the quantities x_{ab} will be small for all transitions with appreciable matrix element, provided the neutron energy is large compared to the level spacing and the neutron velocity large compared to the average nuclear velocities. If these conditions—the latter of which only implies a very weak limitation on the admissible degree of initial excitation of the system—are satisfied, we may therefore expand the integrals in (2.11) in powers of x_{ab} and replace the restricted sum by an unrestricted one. This method will be carried out in the following sections. With appropriate changes it is also applicable to the treatment of other problems. For the scattering of x-rays by atoms, in particular, it leads to certain modifications of hitherto accepted results.[5]

3. AN EXPRESSION FOR THE CROSS SECTION IN TERMS OF AVERAGES OVER THE INITIAL STATE

With the notations $u=4k^2$, $t=\kappa^2$, (2.11) may be written

$$\sigma^{(a)}=\sum_b \sigma_{ab} \tag{3.1}$$

$$\sigma_{ab}=(\sigma_{ab})_+-(\sigma_{ab})_- \tag{3.2}$$

$$(\sigma_{ab})_{\pm}=\frac{4\pi}{u}\int_0^{t_{\pm}}\phi_{ab}(t)dt \tag{3.3}$$

$$t_{\pm}=(u/4)\{1\pm(1-x_{ab})^{\frac{1}{2}}\}^2.$$

The expansion of $(\sigma_{ab})_+$ in powers of x_{ab} is obtained by expanding the integral in powers of $u-t_+$ and expressing this quantity by its expansion in powers of x_{ab}

$$u-t_+=\frac{ux_{ab}}{2}\left\{1+\frac{x_{ab}}{8}+\frac{x_{ab}^2}{16}+\frac{5}{128}x_{ab}^3+\cdots\right\}. \tag{3.4}$$

This yields[6]

$$\frac{(\sigma_{ab})_+}{4\pi}=\frac{1}{u}\int_0^u \phi(t)dt-\frac{x}{2}\phi(u)+\frac{x^2}{16}\{-\phi(u)+2u\phi'(u)\}$$

$$+\frac{x^3}{32}\{-\phi(u)+u\phi'(u)-\tfrac{2}{3}u^2\phi''(u)\}$$

$$+\frac{x^4}{512}\{-10\phi(u)+9u\phi'(u)$$

$$-4u^2\phi''(u)+(4/3)u^3\phi'''(u)\}. \tag{3.5}$$

---

[5] G. Placzek, Bull. Am. Phys. Soc. **27**, No. 1, 13 (1952).
[6] In order to save indices, we write in the following x for x_{ab} and ϕ for ϕ_{ab}.

To expand $(\sigma_{ab})_-$, we expand ϕ in powers of t, integrate and expand t_- in powers of x

$$t_-=\tfrac{1}{16}ux^2(1+\tfrac{1}{2}x+\cdots). \tag{3.6}$$

Noting that $\phi_{ab}(0)=0$ for $a\neq b$ and $x_{ab}=0$ for $a=b$, we obtain

$$\frac{(\sigma_{ab})_-}{4\pi}=\frac{x^4}{512}u\phi'(u)+O(x^5). \tag{3.7}$$

The lower limit of the integrals in (2.11) thus gives rise to a fourth-order correction only.

The quantities σ_{ab} have now to be summed over b. At the same time we may average the cross section over the initial states of the system. Putting

$$\sigma=\sum_a \gamma_a\sigma^{(a)}, \tag{3.8}$$

where γ_a is the probability of finding the system in state "a" before the collision and introducing sums S_n by

$$S_n(t)=\sum_a \gamma_a S_n^{(a)}(t) \tag{3.9}$$

$$S_n^{(a)}(t)=\sum_b^{E_b<E_0+E_a}(E_b-E_a)^n\phi_{ab}(t), \tag{3.10}$$

we obtain for the cross section from (3.1), (3.2), (3.5), (3.7)–(3.10)

$$\frac{\sigma}{4\pi}=\frac{1}{u}\int_0^u S_0(t)dt-\frac{1}{2E_0}S_1(u)+\frac{1}{16E_0^2}\{-S_2(u)+2uS_2'(u)\}$$

$$+\frac{1}{32E_0^3}\{-S_3(u)+uS_3'(u)-\tfrac{2}{3}u^2S_3''(u)\}$$

$$+\frac{1}{512E_0^4}\{-10S_4(u)+u(9S_4'(u)-S_4'(0))$$

$$-4u^2S_4''(u)+(4/3)u^3S_4'''(u)\}. \tag{3.11}$$

We now have to discuss the quantities S_n. With (2.9) we may write (3.10)

$$S_n^{(a)}(\kappa^2)$$

$$=\frac{1}{4\pi}\int d\Omega_\kappa \sum_b^{E_b<E_0+E_a}(E_b-E_a)^n|(F(\kappa))_a{}^b|^2. \tag{3.12}$$

For freely orientable systems such as molecules, the averaging over the directions of κ is superfluous since the sum over b will here depend on the magnitude of κ only. For systems showing the type of random orientation present in polycrystals, however, this averaging process is necessary.

Expressing E_b-E_a as the diagonal element of the operator $H-E_a$

$$E_b-E_a=(H-E_a)_b{}^b$$

and neglecting the limitation of the sum in (3.12), we obtain by closure

$$\sum_b (E_b-E_a)^n |F_a{}^b|^2 = \sum_b (F^*)_b{}^a(\{H-E_a\}^n)_b{}^b F_a{}^b$$

$$= (F^*\{H-E_a\}^n F)_a{}^a. \tag{3.13}$$

From the argument given at the end of the preceding section it may be concluded that under the conditions stated there the error committed by closure will be quite negligible for heavy nuclei and small n. In Sec. VI this will also be shown in somewhat greater detail for a specific example. Combining (3.9), (3.12), and (3.13) we obtain finally

$$S_n(\kappa^2) = \langle F^*(\kappa)(H-E_a)^n F(\kappa)\rangle_{Av}, \tag{3.14}$$

where the average is to be taken (1) over the configuuration of state "a," (2) over the distribution of states "a" initially present in the system, and (3) if necessary, over the directions of κ. The first two averaging processes can sometimes be combined in the customary way by the use of transformation theory.

With the help of the relation,

$$\langle (H-E_a)O\rangle_{Av} = \langle O(H-E_a)\rangle_{Av} = 0,$$

where O is a time-independent operator, (3.14) may be expressed in various ways. A convenient form is

$$\begin{aligned}
S_0 &= (F^*F)_{Av}\\
S_1 &= (F^*[HF])_{Av}\\
S_2 &= ([F^*H][HF])_{Av}\\
S_3 &= ([F^*H][H[HF]])_{Av}\\
S_4 &= ([H[HF^*]][H[HF]])_{Av},
\end{aligned} \tag{3.15}$$

where $[AB]$ is the commutator of the operators A and B. The evaluation of (3.15) will be carried out in Secs. V and VII.

The quantities S_n admit a simple physical interpretation. If a momentum $\hbar\kappa$ is imparted to the system, the average energy transfer accompanying this momentum transfer is given by $S_1(\kappa^2)/S_0(\kappa^2)$ and the n-th moment of the energy transfer by $S_n(\kappa^2)/S_0(\kappa^2)$. This will be used later in connection with the comparison of classical and quantum-mechanical results. The moments so defined are of course different from the moments of the energy transfer occurring in an actual collision. The latter are given by

$$\langle E^l\rangle_{Av} = \sum_a \gamma_a \sum_b (E_b-E_a)^l \sigma_{ab}/\sigma \tag{3.16}$$

and may be expressed in terms of the quantities S_n and their derivatives by noting that the denominator of (3.16) is given by (3.11) while the numerator results from (3.11) by simply replacing S_n by S_{n+l}. The moments (3.16) which are thus, for small n, obtainable by our method, are also simply related to the moments of the energy distribution of the scattered neutrons, given by $\langle (E_0-E)^l\rangle_{Av}$.

4. THE DIFFERENTIAL CROSS SECTION

The treatment of the preceding section can be extended to the differential cross section. The differential cross section associated with the transition $a\to b$ is given by

$$d\sigma_{ab}/d\Omega = (1-x_{ab})^{\frac{1}{2}}\phi_{ab}(\kappa_{ab}^2). \tag{4.1}$$

The factor $(1-x_{ab})^{\frac{1}{2}}$ represents the ratio of the velocities of scattered and incoming neutron and κ_{ab}^2 is determined by the relation,

$$\boldsymbol{\kappa}_{ab} = \mathbf{k}_0 - k_0(1-x_{ab})^{\frac{1}{2}}\boldsymbol{\Omega}, \tag{4.2}$$

where $\boldsymbol{\Omega}$ is a unit vector in the direction of scattering.

For scattering without energy change (4.2) goes over into

$$\boldsymbol{\kappa}_0 = \mathbf{k}_0 - k_0\boldsymbol{\Omega}. \tag{4.3}$$

$\kappa_0{}^2$ is related to the cosine μ_0 of the scattering angle by

$$\mu_0 = 1 - \kappa_0{}^2/2k_0{}^2.$$

From (4.2) and (4.3) we have

$$\kappa_{ab}{}^2 = k_0{}^2(1-x)^{\frac{1}{2}} + k^2(1-(1-x)^{\frac{1}{2}})^2. \tag{4.4}$$

One now expands $\phi_{ab}(\kappa_{ab}^2)$ in powers of $\kappa_{ab}{}^2-\kappa_0{}^2$ and substitutes

$$\kappa_0{}^2 - \kappa_{ab}{}^2 = \tfrac{1}{2}\kappa_0{}^2 x_{ab}$$
$$+ \tfrac{1}{4}(\tfrac{1}{2}\kappa_0{}^2 - k^2)x_{ab}{}^2(1 + \tfrac{1}{2}x_{ab} + \tfrac{5}{16}x_{ab}{}^2 + \cdots).$$

Inserting the resulting expansion for ϕ as well as the expansion of $(1-x_{ab})^{\frac{1}{2}}$ into (4.1), one obtains, after summation over b and averaging over a, the following result, where again $u=4k_0{}^2$, the scattering angle is measured by

$$t = \kappa_0{}^2 = \tfrac{1}{2}u(1-\mu_0) = u\sin^2(\theta/2) \tag{4.5}$$

and S_n stands for $S_n(t)$

$$\frac{d\sigma}{d\Omega} = S_0 - \frac{1}{2E_0}\{S_1 + tS_1'\}$$

$$-\frac{1}{16E_0{}^2}\{2S_2 - (2t+u)S_2' - 2t^2S_2''\}$$

$$-\frac{1}{32E_0{}^3}\{2S_3 - 2tS_3' + tuS_3'' + \tfrac{2}{3}t^3S_3'''\}$$

$$-\frac{1}{512E_0{}^4}\{20S_4 - 2(10t-u)S_4' + (4t^2+4tu-u^2)S_4''$$

$$+ 4t^2(\tfrac{2}{3}t-u)S_4''' - (4/3)t^4S_4''''\}. \tag{4.6}$$

As a check we integrate this expression over $d\Omega$. For this and other purposes it is convenient to write it in the form,

$$\frac{d\sigma}{d\Omega}=S_0-\frac{1}{2E_0}\frac{\partial}{\partial t}tS_1+\frac{1}{16E^2}\frac{\partial}{\partial t}\{(u-2t)S_2+2t^2S_2'\}$$

$$+\frac{1}{32E_0^3}\frac{\partial}{\partial t}\{(u-2t)(S_3-tS_3')-\tfrac{2}{3}t^2S_3''\}$$

$$+\frac{1}{512E_0^4}\frac{\partial}{\partial t}\{(u-2t)(10S_4+(u-10t)S_4'$$

$$+4t^2S_4'')+(4/3)t^4S_4'''\}. \quad (4.7)$$

Introducing (4.7) into

$$\frac{\sigma}{4\pi}=\frac{1}{2}\int_{-1}^{1}\frac{d\sigma}{d\Omega}d\mu_0=\frac{1}{u}\int_0^u\frac{d\sigma}{d\Omega}dt \quad (4.8)$$

one immediately obtains (3.11).

Under actual experimental conditions, the measured differential cross section will depend on the variation of the detector sensitivity with energy. We shall concern ourselves with a $1/v$ detector. In this case the partial differential cross sections have to be multiplied by the ratio of the velocities of incident and scattered neutron before the summation over b. This will just compensate the factor $(1-x_{ab})^{\frac{1}{2}}$ in (4.1). It can be shown that the omission of this factor exempts the factors of S_n, S_n', S_n'', etc. in (4.7) from the operation $\partial/\partial t$. We find thus for the effective differential cross section measured by a $1/v$-detector

$$\left(\frac{d\sigma}{\partial\Omega}\right)_{eff}=S_0-\frac{tS_1'}{2E_0}+\frac{1}{16E_0^2}\{(u-2t)S_2'+2t^2S_2''\}$$

$$+\frac{1}{32E_0^3}\{(u-2t)(S_3'-tS_3'')-\tfrac{2}{3}t^2S_3'''\}$$

$$+\frac{1}{512E_0^4}\{(u-2t)(10S_4'+(u-10t)S_4''$$

$$+4t^2S_4''')+(4/3)t^4S_4''''\}. \quad (4.9)$$

5. SINGLE NUCLEUS

We take as the scattering system a single nucleus in the potential $V(\mathbf{r})$. The scattering amplitude is here given by a single term,

$$F=a_s\exp(iPz/\hbar), \quad (5.1)$$

where $P=\hbar\kappa$ is the momentum transferred to the nucleus and z the component of $\mathbf{r}$ in the direction of $\mathbf{P}$. From (5.1) and (3.15) we have

$$S_0=a_s^2. \quad (5.2)$$

With the Hamiltonian

$$H=p^2/2M+V(\mathbf{r}), \quad (5.3)$$

one obtains for the commutation relations:

$$[HF]=\frac{P}{M}F(\tfrac{1}{2}P+p_z) \quad (5.4)$$

$$[H[HF]]=\frac{P}{M}F\left\{-\frac{\hbar}{i}\frac{\partial V}{\partial z}+\frac{P}{M}(\tfrac{1}{2}P+p_z)^2\right\}, \quad (5.5)$$

where p_z is the component of the momentum of the nucleus in the direction of $\mathbf{P}$.

The expressions (5.4) and (5.5) are now introduced into (3.15). Integration by parts with real eigenfunctions and subsequent averaging over the directions of κ (i.e. $\mathbf{P}$) yields

$$\begin{aligned}
S_1/S_0&=y\\
S_2/S_0&=y^2+(4/3)K_{Av}y\\
S_3/S_0&=y^3+4K_{Av}y^2+B_{Av}y\\
S_4/S_0&=y^4+8K_{Av}y^3+4(\tfrac{4}{5}\langle K^2\rangle_{Av}+B_{Av})y^2+2C_{Av}y,
\end{aligned} \quad (5.6)$$

where

$$y=\hbar^2\kappa^2/2M=P^2/2M. \quad (5.7)$$

K is the kinetic energy of the nucleus and

$$B=\tfrac{1}{3}(\hbar^2/M)\nabla^2V \quad (5.8)$$

$$C=\tfrac{1}{3}(\hbar^2/M)(\mathrm{grad}V)^2. \quad (5.9)$$

It is instructive to compare the expressions (5.6) with their classical value. In classical mechanics the energy transfer accompanying the instantaneous transfer of momentum $\mathbf{P}$ to a particle of momentum $\mathbf{p}$ moving in the potential $V(\mathbf{r})$ is independent of the potential and given by

$$E=(1/2M)\{P^2+2\mathbf{P}\cdot\mathbf{p}\}. \quad (5.10)$$

Averaging the nth power of (5.10) over the momentum distribution and over the directions of $\mathbf{P}$ one obtains

$$S_n/S_0=\frac{1}{n+1}\sum_{l=0}^{[n/2]}4^l\binom{n+1}{2l+1}\langle K^l\rangle_{Av}y^{n-l}, \quad (5.11)$$

where $[n/2]$ stands for the largest integer smaller than or equal to $n/2$. Comparison with (5.6) shows that (5.6) is represented by (5.11) plus additional terms which depend on the potential and which appear for $n>2$ only.

We are now ready for the evaluation of the cross section. Observing that the value y_0 of y corresponding to $\kappa^2=u=4k_0^2$ is

$$y_0=4E_0/\mu \quad (5.12)$$

where $\mu=m/M$, we obtain from (3.11), (5.2), (1.3), and (5.6)

$$\frac{\sigma}{\sigma_s}=1-\frac{2}{\mu}+\frac{1}{\mu^2}\left(3+\frac{1}{3}\frac{\mu K_{Av}}{E_0}\right)-\frac{1}{\mu^3}\left(4+\frac{2}{3}\frac{\mu K_{Av}}{E_0}\right)$$

$$+\frac{1}{\mu^4}\left\{5+\frac{\mu K_{Av}}{E_0}-\frac{1}{32}\frac{\mu^3C_{Av}}{E_0^3}\right\}+\cdots. \quad (5.13)$$

The terms containing $\langle K^2\rangle_{Av}$ and B_{Av} have canceled; one-half of the term containing C_{Av} is contributed by the term $u S_4'(0)$ which arises from the lower limit of the integral in (2.11) and the other half by the terms $S_4(u)$ and $u S_4'(u)$. The terms starting with $1/\mu^n$ represent the terms containing S_n and its derivatives in (3.11). The expansion of the partial cross section σ_{ab} in powers of x_{ab} is thus found to lead to an expansion of the cross section σ in powers of $1/\mu$, where the characteristic energies of the system are measured in terms of E_0/μ. It is, however, important to observe that the quantities $K_{Av}/\mu^{-1}E_0$ and $C_{Av}/(\mu^{-1}E_0)^3$—the former of which represents the square of the ratio of the momenta of nucleus and neutron—may be large. The terms of (5.13) have therefore to be rearranged, by writing

$$\frac{\sigma}{\sigma_s}=\left(1-\frac{2}{\mu}+\frac{3}{\mu^2}+\frac{4}{\mu^3}+\frac{5}{\mu^4}\right)$$
$$+\frac{1}{3}\frac{K_{Av}}{E_0}\left(1-\frac{2}{\mu}+\frac{3}{\mu^2}\right)-\frac{1}{32}\frac{C_{Av}}{\mu E_0^3}. \quad (5.14)$$

Here the first term represents the reduced mass effect, mentioned in Secs. I and II, the second term, of the order of the square of the velocity ratio, the Doppler effect and the third term the effect of the potential.

Introducing the cross section of the free nucleus at rest by

$$\sigma_{free}=\sigma_s(1+1/\mu)^{-2}, \quad (5.15)$$

we have thus

$$\sigma/\sigma_{free}=1+\tfrac{1}{3}K_{Av}/\mu E_0-\tfrac{1}{32}C_{Av}/\mu E_0^3. \quad (5.16)$$

The cross section is thus approximately equal to the cross section of the free nucleus with the same momentum distribution and the direct effect of the binding manifests itself only in the last term, which decreases as E_0^{-3}.

For the average energy transfer one obtains from (3.16) by replacing, in (3.11), S_n by S_{n+1}

$$(E_t)_{Av}=\frac{2}{\mu}\left\{E_0\left(\frac{\mu}{\mu+1}\right)^2-\frac{4}{3}K_{Av}\left(1-\frac{17}{4\mu}+\cdots\right)\right.$$
$$\left.+\frac{B_{Av}}{8E_0}\left(1-\frac{8}{3\mu}+\cdots\right)-\frac{4}{15}\frac{\langle K^2\rangle_{Av}}{\mu E_0}(1+O(1/\mu))\right\}. \quad (5.17)$$

In the higher moments of the energy transfer the direct influence of the binding becomes rapidly more important at low neutron energy. Here a potential term is appreciable in the second moment and leading in the third moment. The energy dependence of these higher moments shows how these quantum effects disappear as the neutron energy increases to a value large compared to μ times the level separation, as can be studied in detail on the basis of the data for an oscillator given in Sec. VI.

For the differential cross section one finds from (5.6) and (4.7)

$$\frac{4\pi}{\sigma_s}\frac{d\sigma}{d\Omega}=1-\frac{4}{\mu}+\frac{1}{\mu^2}\left(2\beta+6\beta^2+\frac{\mu}{3}\frac{K_{Av}}{E_0}\right)$$
$$-\frac{4}{\mu^3}\beta\left\{3\beta+\frac{\mu K_{Av}}{E_0}(1-\beta)\right\}$$
$$+\frac{2}{\mu^4}\left\{\beta^2(3+10\beta-5\beta^2)+\frac{3\mu K_{Av}}{E_0}\beta(1-\beta)\right.$$
$$\left.+\frac{\mu^2 D}{8E_0^2}(1-6\beta+6\beta^2)+\frac{\mu^3 C_{Av}}{64E_0^3}\right\} \quad (5.18)$$

where

$$\beta=\sin^2(\theta/2)$$

and

$$D=\tfrac{4}{5}\langle K^2\rangle_{Av}+B_{Av}.$$

Rearranging terms and introducing (5.15), one has from (5.18)

$$\frac{d\sigma}{d\Omega}=\frac{\sigma_{free}}{4\pi}\left(1+\frac{1}{\mu}\left\{2\cos\theta+\frac{K_{Av}}{3E_0}-\frac{C_{Av}}{32E_0^3}+\cdots\right\}\right.$$
$$+\frac{1}{\mu^2}P_2(\cos\theta)\left\{1+\frac{2}{3}\frac{K_{Av}}{E_0}+\frac{\langle K^2\rangle_{Av}+(5/4)B_{Av}}{5E_0^2}+\cdots\right\}$$
$$+\frac{1}{\mu^3}P_2(\cos\theta)\left\{\frac{K_{Av}}{3E_0}+\cdots\right\}$$
$$\left.+\frac{1}{\mu^4}\left\{\frac{(5\cos^2\theta-1)(1-\cos^2\theta)}{8}+\cdots\right\}\right). \quad (5.19)$$

Equation (5.19) contains two terms depending on the potential which are again small and the remainder is identical with the differential scattering cross section of a free nucleus with the same momentum distribution. Solving the latter problem in the usual way by transforming from center-of-mass to laboratory system and expanding the result in powers of the velocity ratio and the coefficients of this expansion in powers of $1/\mu$, one indeed finds again (5.19) except for the two potential terms.

The relative order of the various terms in (5.19) depends on the neutron energy and the degree of excitation of the system. For high excitation and low neutron energy, for example, it will not be consistent to carry the last two terms of (5.19) since under these conditions they will be smaller than terms which would arise if the original expansion (5.18) were carried to higher order than the fourth. Nevertheless, however, the terms contained in (5.19) are more than amply sufficient for an accurate representation of the differential cross section of heavy nuclei at all neutron energies satisfying the conditions stated at the end of Sec. II.

The effective differential cross section, obtained in the same way as (5.19), turns out to be

$$
\left(\frac{d\sigma}{d\Omega}\right)_{\text{eff}} = 1 + \frac{1}{\mu}\Big\{1 + \cos\theta\Big(1 + \frac{K_{\text{Av}}}{3E_0}
$$
$$
+ \frac{B_{\text{Av}}}{8E_0^2} + \frac{5C_{\text{Av}}}{32E_0^3} + \cdots\Big)\Big\}
$$
$$
+ \frac{\cos\theta}{\mu^2}\Big\{1 + \frac{2K_{\text{Av}}}{3\,E_0} + \frac{B_{\text{Av}}+(2/5)\langle K^2\rangle_{\text{Av}}}{2E_0^2} + \cdots\Big\}
$$
$$
+ \frac{\cos\theta}{2\mu^3}\Big\{\sin^2\theta + (11 - 9\cos^2\theta)\frac{K_{\text{Av}}}{3E_0} + \cdots\Big\}
$$
$$
+ \frac{\cos\theta}{2\mu^4}\{\sin^2\theta + \cdots\}. \tag{5.20}
$$

From the results of this section it appears that, while the collision is describable in terms of the classical particle picture only if the neutron energy E_0 is large compared to μ times the level separation Δ, the quantum characteristics of the collision will, in the energy region $\Delta \ll E_0 \ll \mu\Delta$, manifest themselves primarily in the energy distribution of the scattered neutrons. The higher moments of the energy transfer show a strong explicit dependence on the binding potential. In the cross section and the differential cross sections, on the other hand, this dependence is much less pronounced and the binding affects these quantities mainly through its influence on the nuclear momentum distribution.

6. ISOTROPIC OSCILLATOR

As an example we consider a three-dimensional isotropic harmonic oscillator of frequency ω at a temperature T. For this system one finds easily

$$
K_{\text{Av}} = \tfrac{3}{2}T_{\text{eff}} \qquad B = (\hbar\omega)^2
$$
$$
\langle K^2\rangle_{\text{Av}} = (15/4)T_{\text{eff}}^2 \qquad C_{\text{Av}} = \tfrac{2}{3}BK_{\text{Av}} = (\hbar\omega)^2 T_{\text{eff}}, \tag{6.1}
$$

where the effective temperature T_{eff} is related to the temperature T by

$$
T_{\text{eff}}/T = (\hbar\omega/2T)\coth(\hbar\omega/2T). \tag{6.2}
$$

For (5.16) we have thus for the cross section

$$
\frac{\sigma}{\sigma_{\text{free}}} = 1 + \frac{T_{\text{eff}}}{2\mu E_0}\Big\{1 - \frac{1}{16}\Big(\frac{\hbar\omega}{E_0}\Big)^2\Big\}. \tag{6.3}
$$

The corresponding expressions for the differential cross sections and the moments of the energy transfer follow directly by insertion of (6.1) into the general formulas of the preceding section and will thus not be written down here.

For $T = 0$, (6.3) goes over into

$$
\frac{\sigma}{\sigma_{\text{free}}} = 1 + \frac{1}{4\mu n_0}\Big(1 - \frac{1}{16n_0^2}\Big), \tag{6.4}
$$

where $n_0 = E_0/\hbar\omega$.

The asymptotic expression (6.4), valid for large n_0, may now be compared with the exact one. The transition probabilities are given by[2]

$$
\phi_{0n}(\kappa^2) = a_s^2\xi^n \exp(-\xi)/n! \tag{6.5}
$$
$$
\xi = \hbar\kappa^2/2M\omega = P^2/2M\hbar\omega = y/\hbar\omega. \tag{6.5a}
$$

Introduction of (6.5) into (3.1)–(3.3) yields

$$
\frac{\sigma}{\sigma_{\text{free}}} = \frac{(\mu+1)^2}{4\mu n_0}\sum_{n=0}^{[n_0]}\{\Pi_n(\xi_n^{(-)}) - \Pi_n(\xi_n^{(+)})\}, \tag{6.6}
$$

where

$$
\xi_n^{(\pm)} = (2/\mu)\{n_0 - \tfrac{1}{2}n \pm [n_0(n_0 - n)]^{\frac{1}{2}}\}
$$
$$
\Pi_n(x) = \frac{1}{n!}\int_x^\infty \xi^n e^{-\xi}d\xi = e^{-x}\sum_{l=0}^{n}\frac{x^l}{l!}
$$
$$
= 1 - \frac{x^{n+1}e^{-x}}{(n+1)!}\sum_{m=0}^{\infty}\frac{(n+1)!}{(n+m+1)!}x^m. \tag{6.7}
$$

Since for large μ the variation of the cross section with energy is a small effect, it is convenient to write the cross section in the form,

$$
\frac{\sigma}{\sigma_{\text{free}}} = 1 + \frac{c(n_0)}{\mu n_0} \tag{6.8}
$$

and discuss $c(n_0)$ rather than σ.

From (6.4) we have then

$$
c(n_0) = \tfrac{1}{4}\{1 - 1/(4n_0)^2\} \tag{6.9}
$$

while from (6.6)

$$
c(n_0) = -\mu n_0 + \frac{(\mu+1)^2}{2}\sum_{n=0}^{[n_0]}\{\Pi_n(\xi_n^-) - \Pi_n(\xi_n^+)\}. \tag{6.10}
$$

Numerical comparison of these two expressions (see Fig. 1) shows, that for $\mu = 12$ their difference becomes negligible at $n_0 \approx 4.2$ and in the limit of large μ already at $n_0 \approx 2$. Analytically, this may be seen as follows: Expanding (6.10) in powers of $1/\mu$,

$$
c = c_0 + c_1/\mu + c_2/\mu^2 + \cdots, \tag{6.11}
$$

one finds

$$
c_0 = \begin{cases} (2n_0-1)(n_0(n_0-1))^{\frac{1}{2}} - 2n_0(n_0-1) & \text{for } n_0 > 1 \\ 2n_0(1 - n_0) & \text{for } n_0 < 1. \end{cases} \tag{6.12}
$$

By expansion in powers of $1/n_0$, (6.12) goes over into (6.9). The quantities c_n/μ^n contribute asymptotically only to terms of higher order than (6.9) but reach their asymptotic form for $n_0 \gg n$ only. Consequently, the deviations of (6.10) from (6.9) will extend to higher and higher n_0 as μ decreases. For $\mu = 1$, $c(n_0)$ becomes, for large n_0, a periodic function of n_0, with period 1 and average $\tfrac{1}{4}$.[3,7]

---

[7] A. Messiah, J. phys. et radium **12**, 670 (1951).

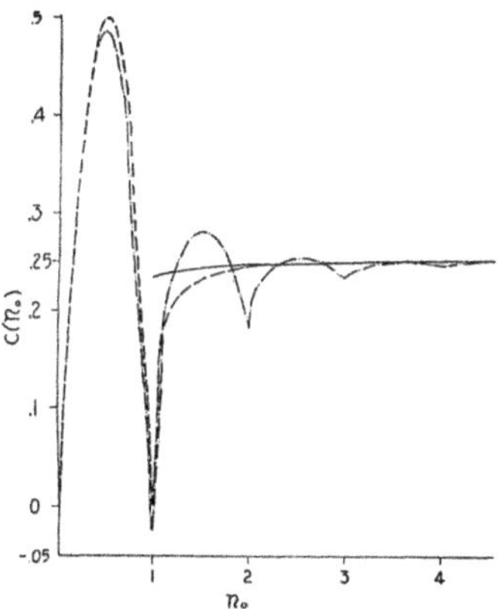

FIG. 1. The coefficient $c(n_0)$ in Eq. (6.8) for the cross section of isotropic oscillator at zero temperature; ——— asymptotic, Eq. (6.9); —·—·—·—·— $\mu=12$, Eq. (6.10); — — — — — $\mu=\infty$, Eq. (6.12).

It may also be noted that for the ground state of the oscillator the sums S_l can be evaluated explicitly. From (6.5) and (3.10) we have for the restricted sums

$$S_l(\xi, n_0) = a_s^2 e^{-\xi} \sum_{n}^{[n_0]} (n^l/n!)\xi^n, \tag{6.13}$$

where ξ is related to κ^2 and y by (6.5a).

For $l=0$ (6.13) becomes with (6.7)

$$S_0(\xi, n_0) = a_s^2 \amalg_{[n_0]}(\xi). \tag{6.14}$$

S_l may be expressed in terms of S_0 by decomposing n^l into binomial coefficients. In this way one finds

$$S_1(\xi, n_0) = \xi S_0(\xi, n_0-1)$$
$$S_2(\xi, n_0) = \xi^2 S_0(\xi, n_0-2) + \xi S_0(\xi, n_0-1)$$
$$S_3(\xi, n_0) = \xi^3 S_0(\xi, n_0-3) + 3\xi^2 S_0(\xi, n_0-2) + \xi S_0(\xi, n_0-1) \tag{6.15}$$
$$S_4(\xi, n_0) = \xi^4 S_0(\xi, n_0-4) + 6\xi^3 S_0(\xi, n_0-3)$$
$$+7\xi^2 S_0(\xi, n_0-2) + \xi S_0(\xi, n_0-1).$$

If $S_0(\xi, n_0-j)$ is replaced by $S_0(\xi, \infty) = a_s^2$, (6.15) goes over into (5.6) which can also be checked by introducing (6.1) into (5.6). The derivatives of S_l with respect to ξ may also be expressed in terms of S_0 with the aid of the relation,

$$\xi S_l' = S_{l+1} - \xi S_0. \tag{6.16}$$

From (6.14) we have for the relative error in S_0 caused by the closure approximation

$$\delta_0 = \frac{S_0(\xi, \infty) - S_0(\xi, n_0)}{S_0(\xi, \infty)} = 1 - \amalg_{[n_0]}(4\beta n_0/\mu), \tag{6.17}$$

where

$$\beta = \kappa^2/4k_0^2 = \mu\xi/4n_0 < 1.$$

If $4\beta/\mu$ is small compared to one, which for large μ will be true for all values of β, (6.17) goes over into

$$\delta_0 = \frac{1}{([n_0]+1)!}\left(\frac{4\beta n_0}{\mu}\right)^{[n_0]+1}\exp(-4\beta n_0/\mu). \tag{6.18}$$

For large n_0 we have, expressing $[n_0]$ by

$$[n_0] = n_0 - 1 + \epsilon,$$

where ϵ varies between 0 and 1 and using Stirlings' formula,

$$\delta_0 = (2\pi n_0)^{-\frac{1}{2}}(4\beta/\mu)^{\epsilon}\{(4\beta/\mu)\exp(1-4\beta/\mu)\}^{n_0}. \tag{6.19}$$

If μ is large, δ_0 will thus decrease exponentially with increasing energy for all values of β.

If $4\beta/\mu$ is no longer small, we have to revert to the original expression (6.17). From the integral representation of $\amalg$ it is then seen that for $4\beta/\mu \geq 1$, δ_0 does not vanish for large n_0, but tends to the limit $\frac{1}{2}$. A more detailed analysis, taking compensation effects into account, shows, however, that the closure correction to $c(n_0)$ decreases for all $\mu > 1$ as $\exp(-\alpha n_0)$ where

$$\alpha = -q - \log(1-q), \quad q = [(\mu-1)/(\mu+1)]^2. \tag{6.20}$$
$$\text{For} \quad 1-q \ll 1, \quad \alpha = \log(4/\mu) - 1,$$
$$\text{for} \quad q \ll 1, \quad \alpha = \tfrac{1}{2}[(\mu-1)/(\mu+1)]^4.$$

7. SEVERAL NUCLEI

We now have to generalize the results of Sec. V to a system of N nuclei with the Hamiltonian

$$H = V(\mathbf{r}_1, \mathbf{r}_2, \cdots \mathbf{r}_N) + \sum_s p_s^2/2M_s. \tag{7.1}$$

At this stage it becomes necessary to consider the spin dependence of the scattering lengths. For a nucleus of spin j_s, the scattering length is given by[8]

$$a_s = \frac{1}{2j_s+1}\{(j_s+1)a_s^{(+)} + j_s a_s^{(-)}$$
$$+ 2(\boldsymbol{\alpha}_n \cdot \boldsymbol{\alpha}_s)(a_s^{(+)} - a_s^{(-)})\}. \tag{7.2}$$

$a_s^{(+)}$ and $a_s^{(-)}$ are the scattering length associated with scattering processes leading to final states of the system nucleus plus neutron with spin $j_s+\frac{1}{2}$ and $j_s-\frac{1}{2}$, respectively, and $\boldsymbol{\alpha}_s$ and $\boldsymbol{\alpha}_n$ are the Pauli spin operators for nucleus and neutron.

Averaging over the orientations of the neutron spin, we have, with

$$\langle \boldsymbol{\alpha}_n \cdot \boldsymbol{\alpha}_s \rangle_{Av} = 0; \quad \langle (\boldsymbol{\alpha}_n \cdot \boldsymbol{\alpha}_s)^2 \rangle_{Av} = \tfrac{1}{4}j_s(j_s+1), \tag{7.3}$$

for the scattering cross section of the bound nucleus

$$\frac{\sigma_s}{4\pi} = \langle a_s^2 \rangle_{Av} = \frac{1}{2j_s+1}\{(j_s+1)(a_s^{(+)})^2 + j_s(a_s^{(-)})^2\}. \tag{7.4}$$

The definition of the matrix element (1.4) has now to be modified by inclusion of the spin coordinates. If the spins of different nuclei are not correlated—and this excludes homoeonuclear molecules at low temperatures, the averaging of (3.12) over position and spin coordinates may be carried out independently. As a consequence, the product $a_s a_{s'}$ has to be replaced by

$$\langle a_s a_{s'} \rangle_{Av} = \begin{cases} \langle a_s \rangle_{Av}\langle a_{s'} \rangle_{Av} = a_s^{(c)} a_{s'}^{(c)} & \text{for} \quad s' \neq s \\ \langle a_s^2 \rangle_{Av} = \sigma_s/4\pi & \text{for} \quad s' = s, \end{cases} \tag{7.5}$$

where the coherent scattering length $a_s^{(c)}$ is defined by

$$a_s^{(c)} = \langle a_s \rangle_{Av} = \frac{1}{2j_s+1}\{(j_s+1)a_s^{(+)} + j_s a_s^{(-)}\} \tag{7.6}$$

and $\langle a_s^2 \rangle_{Av}$ is given by (7.4).

[8] M. Hamermesh and J. Schwinger, Phys. Rev. **69**, 145 (1945).

The cross section (1.1) and the differential cross sections may then be written as a sum of two parts which are usually denoted as the coherent and the incoherent cross section. The coherent cross section is obtained by replacing a_s with $a_s^{(c)}$ throughout, while in the incoherent cross section the product terms with $s' \neq s$ are omitted and a_s^2 is replaced by

$$\langle a_s^2 \rangle_{Av} - (\langle a_s \rangle_{Av})^2 = \frac{j_s(j_s+1)}{2j_s+1}(a_s^{(+)} - a_s^{(-)})^2.$$

The expressions of Sec. V for the total and the differential cross sections of a system containing a single nucleus will thus directly represent the incoherent cross section of a system of several nuclei with uncorrelated spins, if σ_{free} is replaced by the incoherent cross section $\sigma_{f_s}^{(i)}$ of a free nucleus at rest

$$\sigma_{f_s}^{(i)} = 4\pi \left(\frac{\mu_s}{\mu_s+1}\right)^2 \frac{j_s(j_s+1)}{2j_s+1}(a_s^{(+)} - a_s^{(-)})^2. \quad (7.7)$$

The averages entering these expressions are then to be taken for nucleus s and, in particular, the differentiations in the definitions (5.8) and (5.9) of B and C with respect to the coordinates of nucleus s. For the total incoherent scattering cross section we have thus

$$\sigma^{(i)} = \sum_s \sigma_{f_s}^{(i)} \left\{ 1 + \frac{1}{3}\frac{\langle \dot{K}_s \rangle_{Av}}{\mu_s E_0} - \frac{1}{32}\frac{\langle C_s \rangle_{Av}}{\mu_s E_0^3} \right\}. \quad (7.8)$$

As an example let us consider a Debye crystal containing a single type of nuclei. One has then

$$K_{Av} = \frac{36T^4}{\Theta^3} \int_0^{\Theta/2T} x^3 \coth x\, dx$$

$$B_{Av} = \tfrac{3}{5}\Theta^2$$

$$C_{Av} = \frac{96T^6}{\Theta^3} \int_0^{\Theta/2T} x^5 \coth x\, dx \qquad (7.9)$$

$$\langle K^2 \rangle_{Av} = (5/3)(K_{Av})^2,$$

where Θ is the Debye temperature.

For $T \ll \Theta$

$$K_{Av} = \frac{9}{16}\Theta \left\{ 1 + \frac{8}{15}\left(\frac{\pi T}{\Theta}\right)^4 \right\}$$

$$C_{Av} = \frac{1}{4}\Theta^3 \left\{ 1 + \frac{32}{21}\left(\frac{\pi T}{\Theta}\right)^6 \right\} \qquad (7.10)$$

for $T \gg \Theta$

$$K_{Av} = \frac{3}{2}T \left\{ 1 + \frac{1}{20}\left(\frac{\Theta}{T}\right)^2 \right\}$$

$$C_{Av} = \frac{3}{5}T\Theta^2 \left\{ 1 + \frac{5}{84}\left(\frac{\Theta}{T}\right)^2 \right\}. \qquad (7.11)$$

The incoherent cross section per nucleus becomes then

$$\frac{\sigma^{(i)}}{\sigma_{free}^{(i)}} = 1 + \frac{K_{Av}}{3\mu E_0}\left\{ 1 - \frac{\nu}{24}\left(\frac{\Theta}{E_0}\right)^2 \right\}. \quad (7.12)$$

The coefficient ν depends but little on the temperature; it decreases from $\nu = 1$ for $T \ll \Theta$ to $\nu = 9/10$ for $T \gg \Theta$. For heavy nuclei (7.12) will hold as soon as the neutron energy is slightly larger than the Debye temperature. In the effective differential cross section the direct binding effect is somewhat larger as a result of the presence, in the first line at (5.20), of the term containing B_{Av} which at low temperatures and $E_0 = 2\Theta$ amounts to twenty percent of the Doppler term.[9]

Returning now to the general problem it will be convenient to abandon the customary division of the cross section into a coherent and an incoherent part. Instead, we shall distinguish between diagonal terms $(s = s')$ and interference terms $(s \neq s')$. The diagonal terms are represented by the results of Sec. V,[10] so that we only have to calculate the interference terms. For this purpose we express the quantities S_n by

$$S_n = \sum_s \langle a_s^2 \rangle_{Av} G_n^{(ss)} + {\sum}'_{ss'} a_s^{(c)} a_{s'}^{(c)} G_n^{(ss')}. \quad (7.13)$$

$G_n^{(ss)}$ is given by (5.6). For $G_n^{(ss')}$ we have from (3.12) with the notation

$$f_s = \exp(i\kappa \cdot \mathbf{r}_s) = \exp(iPz_s/\hbar),$$

$$G_0^{(ss')} = \langle f_{s'}^* f_s \rangle_{Av} = \left\langle \frac{\sin \kappa r_{ss'}}{\kappa r_{ss'}} \right\rangle_{Av}$$

$$G_1^{(ss')} = \langle f_{s'}^*[Hf_s] \rangle_{Av} \qquad (7.14)$$

$$G_2^{(ss')} = \langle [f_{s'}^* H][Hf_s] \rangle_{Av}$$

$$G_3^{(ss')} = \langle [f_{s'}^* H][H[Hf_s]] \rangle_{Av}.$$

[9] The statements made so far on the basis of the usual theory of neutron scattering by crystals [J. M. Cassels, *Progress in Nuclear Physics*, Vol. I (edited by O. R. Frisch, London-New York, 1950)] about the behavior of the cross section at high energies are either entirely erroneous because of inconsistent approximations [R. Weinstock, Phys. Rev. **65**, 1 (1944); R. J. Finkelstein, Phys. Rev. **72**, 907 (1947)], or very incomplete. [A. Akhiezer and I. Pomeranchuk, J. Phys. (U.S.S.R.) **11**, 167 (1947); D. A. Kleinman, thesis, Brown University (1951).] Akhiezer and Pomeranchuk have given an integral representation of the cross section. From its discussion they conclude that the neutron energy, at which the direct binding effects in the cross section become negligible, is determined by a condition which is essentially identical with Eq. (2.2) rather than (2.3) and which is therefore much too restrictive. Furthermore, their treatment leads to the same energy dependence for coherent and incoherent cross section at high energy. Actually, however, the derivatives of these two quantities with respect to the energy have opposite sign at high energy. The evaluation of the asymptotic interference term (7.21) for crystals [Placzek, Nijboer, and Van Hove, Phys. Rev. **82**, 392 (1951)] shows that this term, which has negative sign, is for heavy nuclei of considerably larger size than the Doppler term. At high energy therefore the coherent cross section increases with increasing energy, in contrast to the incoherent cross section which decreases with increasing energy.

[10] Adding the diagonal terms of the coherent cross section to the expressions (7.8) for the incoherent cross section is equivalent to replacing $\sigma_{f_s}^{(i)}$ by the total free cross section σ_{f_s}.

386 G. PLACZEK

The commutation relations of f_s with the Hamiltonian (7.1) are identical with (5.4)–(5.5),

$$[Hf_s] = \frac{P}{M_s} f_s(\tfrac{1}{2}P + p_{sz}) \tag{7.15}$$

$$[H[Hf_s]] = \frac{P}{M_s} f_s\left\{ -\frac{h}{i}\frac{\partial V}{\partial z_s} + \frac{P}{M_s}(\tfrac{1}{2}P + p_{sz})^2 \right\}. \tag{7.16}$$

With the aid of the relation

$$\langle f_s f_{s'}{}^*(\tfrac{1}{2}P + p_{sz})\rangle_{\mathrm{Av}} = \langle f_s f_{s'}{}^*(\tfrac{1}{2}P - p_{s'z})\rangle_{\mathrm{Av}} = 0 \quad \text{for} \quad s \neq s'$$

obtained through integration by parts, one finds from (7.14) and (7.15):

$$G_1^{(ss')} = 0 \tag{7.17}$$

$$G_2^{(ss')} = \frac{P^4}{4M_s M_{s'}} G_0^{(ss')}$$
$$+ \frac{h^2}{M_s M_{s'}} \langle \exp(i\boldsymbol{\kappa}\cdot\mathbf{r}_{ss'})(\mathbf{p}_s\cdot\boldsymbol{\kappa})(\mathbf{p}_{s'}\cdot\boldsymbol{\kappa})\rangle_{\mathrm{Av}}. \tag{7.18}$$

From (7.17) it follows that the first-order reduced mass correction does not apply to the interference terms. The second term in (7.18) is the analog of the Doppler term encountered previously and represents an effect caused by the correlation of the momenta of different nuclei. In classical statistics there is no such correlation and the term will thus vanish at high temperatures.

In a diatomic molecule, for example, the translational momenta of the two nuclei are fully correlated and the momenta of relative motion fully anticorrelated. At high temperatures, when the relative motion is fully excited, these two effects cancel out while at low temperatures the negative correlation prevails. The result of the evaluation of the term for a particular case will be given in the discussion of the cross sections; now let us just consider the average of the product of the momenta alone.

Expressing the momenta $\mathbf{p}_1$ and $\mathbf{p}_2$ of the nuclei in a diatomic molecule by the momentum $\mathbf{p}_c$ of the center of gravity and the momentum $\mathbf{p}$ of relative motion, we have

$$\mathbf{p}_1 = \frac{M_1}{M_1 + M_2}\mathbf{p}_c + \mathbf{p}, \quad \mathbf{p}_2 = \frac{M_2}{M_1 + M_2}\mathbf{p}_c - \mathbf{p},$$

$$(\mathbf{p}_1\cdot\mathbf{p}_2)_{\mathrm{Av}} = \frac{M_1 M_2}{(M_1 + M_2)^2}\langle p_c{}^2\rangle_{\mathrm{Av}} - \langle p^2\rangle_{\mathrm{Av}} = 2M_r(E_t - E_i),$$

where E_t and E_i are the average kinetic energies of translation and internal motion. At high temperatures $E_i = E_t = \tfrac{3}{2}T$; at low temperatures, $T \ll \hbar\omega$, $E_t - E_i = -\hbar\omega/4 + T/2$ or $-\hbar\omega/4 + 3T/2$ according to whether the temperature is large or small compared to the rotational quanta.

For G_3 one finds

$$G_3^{(ss')} = \frac{P^2}{M_s}G_2^{(ss')} + \frac{P^2}{M_s M_{s'}}\left\langle f_s f_{s'}{}^*\left\{ \frac{\hbar^2}{2}\frac{\partial^2 V}{\partial z_s \partial z_{s'}} \right.\right.$$
$$\left.\left. + \frac{P}{M_s}p_{sz}{}^2(p_{s'z} - \tfrac{1}{2}P) \right\} \right\rangle_{\mathrm{Av}}. \tag{7.19}$$

The expression for G_4 is longer and will not be given here. The discussion of the higher approximations in the preceding sections was necessary mainly to gain clear insight into the structure of the theory. This having been achieved we can limit ourselves, in the discussion of the interference effects, to the terms represented by S_0, S_1, and S_2. In this approximation one finds from (3.11) for the contribution of the interference terms to the cross section

$$\sigma_{\mathrm{int}} = 8\pi\sum_{s,\,s'}^{s'<s} a_s^{(c)}a_{s'}^{(c)}\left\{ \frac{1}{2k_0{}^2}\left\langle\frac{1}{r_{ss'}{}^2}\right\rangle_{\mathrm{Av}} - \frac{1}{2k_0{}^2}\left\langle\frac{\cos 2k_0 r_{ss'}}{r_{ss'}{}^2}\right\rangle_{\mathrm{Av}} \right.$$
$$+ \frac{\langle 2k_0 z_{ss'}\sin 2k_0 z_{ss'}v_{sz}v_{s'z}\rangle_{\mathrm{Av}}}{v_0{}^2}$$
$$\left. + \frac{1}{\mu_s\mu_{s'}}\left\langle\frac{\sin 2k_0 r_{ss'}}{2k_0 r_{ss'}} + \cos 2k_0 r_{ss'}\right\rangle_{\mathrm{Av}} \right\}. \tag{7.20}$$

Here v_0 denotes the neutron velocity, and v_{sz} and $v_{s'z}$ components of the nuclear velocities; $z_{ss'}$ stands for $\mathbf{r}_{ss'}\cdot\mathbf{k}_0/k_0$ and v_{sz} for $\mathbf{v}_s\cdot\mathbf{k}_0/k_0$. It will of course be kept in mind that the quantities v_{sz} and $v_{s'z}$ do not commute with $z_{ss'}$. The first two terms in (7.20) represent the static approximation and the rest the corrections resulting from G_2. At high temperatures the third term vanishes. For short neutron wavelength all the terms except the first one are not only rapidly fluctuating but also their magnitude goes to zero with decreasing wavelength.[11] The cross section is then represented by

$$\sigma_{\mathrm{int}} = \frac{4\pi}{k_0{}^2}\sum_{s,\,s'}^{s'<s} a_s^{(c)}a_{s'}^{(c)}\langle r_{ss'}{}^{-2}\rangle_{\mathrm{Av}}. \tag{7.21}$$

The wavelength at which (7.20) goes over into (7.21) depends upon the nature of the scattering system and its determination requires rather careful considerations.[12]

Introducing (7.13) and (7.18) into (4.7) or (4.9), one obtains simple but somewhat lengthy expressions for the contribution of the interference terms to the differential cross section. In an abbreviated form they may be written as follows:

$$\left(\frac{d\sigma_{\mathrm{int}}}{d\Omega}\right)_{\mathrm{eff}} = \varphi_0 + \left\{ (1 - 2\beta)\frac{\partial}{\partial\beta} \right.$$
$$\left. + 2\beta^2\frac{\partial^2}{\partial\beta^2} \right\}(\beta\varphi_1 + \beta^2\varphi_2) \tag{7.22}$$

$$\frac{d\sigma_{\mathrm{int}}}{d\Omega} = \left(\frac{d\sigma_{\mathrm{int}}}{d\Omega}\right)_{\mathrm{eff}} + 2\left(-1 + 2\beta\frac{\partial}{\partial\beta}\right)(\beta\varphi_1 + \beta^2\varphi_2), \tag{7.23}$$

---

[11] This also holds for systems with long-range order; see reference 12.

[12] Placzek, Nijboer, and Van Hove, Phys. Rev. **82**, 392 (1951).

where $\beta = \sin^2\tfrac{1}{2}\theta$. The quantities φ_n are functions of $\kappa_0^2 = 4k_0^2\beta$

$$\varphi_0 = 2 \sum_{s,\,s'}^{s'<s} a_s^{(c)} a_{s'}^{(c)} \left\langle \frac{\sin\kappa_0 r_{ss'}}{\kappa_0 r_{ss'}} \right\rangle_{\mathrm{Av}}$$

$$\varphi_1 = 2 \sum_{s,\,s'}^{s'<s} a_s^{(c)} a_{s'}^{(c)} \langle \cos\kappa_0 z_{ss'} v_{sz} v_{s'z} \rangle_{\mathrm{Av}} / v_0^2$$

$$\varphi_2 = 2 \sum_{s,\,s'}^{s'<s} \frac{a_s^{(c)}}{\mu_s} \frac{a_{s'}^{(c)}}{\mu_{s'}} \left\langle \frac{\sin\kappa_0 r_{ss'}}{\kappa_0 r_{ss'}} \right\rangle_{\mathrm{Av}}.$$

For a diatomic molecule the differential cross sections may be expressed in terms of functions simply related to the error function of complex argument; if, in particular, the effective wavelength $1/\kappa_0$ is large compared to the vibrational amplitude the evaluation of φ_0 and φ_2 is trivial and for φ_1 one then finds for temperatures large compared to the rotational quanta

$$\varphi_1 = 2a_1^{(c)} a_2^{(c)} \langle (\cos\kappa_0 z_{12}) v_{1z} v_{2z} \rangle_{\mathrm{Av}} / v_0^2 =$$
$$- a_1^{(c)} a_2^{(c)} \frac{(E_{\mathrm{vib}} - T)}{3(\mu_1 + \mu_2) E_0} \left(\frac{\pi}{2\kappa_0 r_0} \right)^{\tfrac{1}{2}}$$
$$\times \{ J_{1/2}(\kappa_0 r_0) - 2J_{5/2}(\kappa_0 r_0) \}. \quad (7.24)$$

Here r_0 is the equilibrium distance of the nuclei and E_{vib} the average total vibrational energy. (7.24) provides a simple example for the structure of momentum correlation term. For $T \gg \hbar\omega$, $E_{\mathrm{vib}} - T = 0$ and φ_1 vanishes.

8. APPLICATIONS

The isolation of small electronic contributions to the scattering cross section, which may be caused by ordinary magnetic effects,[13,14] scattering by spin waves[15] and the spin-independent interaction between neutron and electron[16,14] requires a very precise theoretical determination of the dependence of the nuclear scattering on neutron energy or scattering angle. Off hand this may seem too pretentious a task if the scattering system is as complex as in the transmission experiments of Rainwater, Rabi, and Havens on liquid bismuth.[17] On the basis of the results derived above, however, it can be approached with reasonable assurance. The characteristic temperature of liquid Bi, corresponding to the Debye temperature of a solid, is of the order of 100 degrees abs. The expressions (5.16) and (7.20) for the cross section will thus be valid for neutron energies large compared to 10^{-2} ev. Since

$\mu = 209$, the last term in (7.20), which is quadratic in the mass ratio, may be neglected. Since the melting temperature is 544 degrees abs and thus large compared to the characteristic temperature, the effect of the correlation of the momenta, given by the third term in (7.20) disappears, while in the second term of (5.16) $K_{\mathrm{Av}} = 3T/2$. With a slight modification[18,12] of the first two terms of (7.20) which is equivalent to neglecting deflections by an angle $k_0^2 d^2$, where d is of the order of the total linear dimensions of the scattering system, the scattering cross section per nucleus is thus given by

$$\sigma = \sigma_{\mathrm{free}} \left\{ 1 + \frac{T}{2\mu E_0} \left(1 - \frac{C_{\mathrm{Av}}}{16 T E_0^2} \right) \right\}$$
$$- \sigma_{\mathrm{coh}} \frac{2\pi}{k_0^2} \int_0^\infty (1 - \cos 2k_0 r)\{\rho - g(r)\} dr. \quad (8.1)$$

Here σ_{free} is, as before, the total scattering cross section of the free nucleus at rest and $\sigma_{\mathrm{coh}} = 4\pi(a_s^{(c)})^2$ the coherent cross section of the bound nucleus; $g(r)$ is the density at distance r from a given nucleus and ρ the ordinary density. The evaluation of the interference term in (8.1) has been carried out by Placzek, Nijboer, and Van Hove.[12] The third term in (8.1) depends on the average square of the force acting on the nucleus which cannot be calculated precisely for a liquid. Even the crudest estimates, however, are sufficient to show that at the relevant neutron energies this term cannot amount to more than a fraction of the Doppler term.

The discussion of the electronic contributions to the cross section has, in part, been given previously.[14,16] For an isolated rare gas atom paramagnetic scattering is absent and diamagnetic scattering negligible.[14] For liquid bismuth, on the other hand, the ordinary magnetic effects are hardly accessible to a satisfactory theoretical analysis. Although they do not interfere with the nuclear scattering the possibility that they might be of relevant size can by no means be excluded *a priori*. Since they are entirely caused by the outer electrons, however, their energy dependence will differ from that of the contribution of the spin-independent neutron-electron interaction which comes from all the electrons. The energy dependence of the latter has been calculated on the basis of the ordinary form factor.[16] Its isolation would thus seem to require a study of the variation with energy of the difference between the observed cross section and (8.1).

Among other applications of results derived here, the problem of the energy dependence and angular distribution of neutron scattering by heavy molecules might in particular be mentioned. The theory in the form developed here is valid for neutron energies large compared to the vibrational quanta, while many of the

[13] O. Halpern and M. H. Johnson, Phys. Rev. **55**, 898 (1939); O. Halpern, Phys. Rev. **72**, 746 (1947).

[14] E. Fermi and L. Marshall, Phys. Rev. **72**, 1139 (1947).

[15] R. G. Moorhouse, Proc. Phys. Soc. (London), A**64**, 207, 1097 (1951).

[16] Havens, Rainwater, and Rabi, Phys. Rev. **72**, 634 (1947).

[17] Rainwater, Rabi, and Havens, Phys. Rev. **75**, 1295 (1949); **82**, 345 (1951).

[18] F. Zernike and J. A. Prins, Z. Physik **41**, 184 (1927).

388   G. PLACZEK

experiments[19,20] carried out so far concern neutron energies large compared to the rotational but small compared to the vibrational quanta. In order to extend the theory to this region the Hamiltonian (7.1) is replaced by the Hamiltonian of a rigid molecule. The results for the diagonal terms obtained in this way agree for high temperatures with the semiclassical mass tensor approximation[21] and differ from it for low temperatures to which it is not directly applicable. In addition one obtains the interference terms which are not given by the mass tensor approximation and which are particularly important for the angular distribution. The discussion of these results and their comparison with the experiments and with the approximation of Alcock and Hurst[20] will be given in a separate paper.

This paper is based on work begun at the General Electric Research Laboratory and carried out at The Institute for Advanced Study. I am indebted to H. A. Bethe for helpful discussions and to G. F. Chew and R. Jost for a critical reading of the manuscript.

[19] E. Melkonian, Phys. Rev. **76**, 1744 (1949).
[20] N. Z. Alcock and D. G. Hurst, Phys. Rev. **75**, 1609 (1949); **83**, 1100 (1951).
[21] R. G. Sachs and E. Teller, Phys. Rev. **60**, 18 (1941).